Recent Titles in This Series

160 **K. Nomizu, Editor,** Selected Papers on Number Theory, Algebraic Geometry, and Differential Geometry
159 **O. A. Ladyzhenskaya, Editor,** Proceedings of the St. Petersburg Mathematical Society, Volume II
158 **A. K. Kelmans, Editor,** Selected Topics in Discrete Mathematics: Proceedings of the Moscow Discrete Mathematics Seminar, 1972–1990
157 **M. Sh. Birman, Editor,** Wave Propagation. Scattering Theory
156 **V. N. Gerasimov, N. G. Nesterenko, and A. I. Valitskas,** Three Papers on Algebras and Their Representations
155 **O. A. Ladyzhenskaya and A. M. Vershik, Editors,** Proceedings of the St. Petersburg Mathematical Society, Volume I
154 **V. A. Artamonov et al.,** Selected Papers in K-Theory
153 **S. G. Gindikin, Editor,** Singularity Theory and Some Problems of Functional Analysis
152 **H. Draškovičová et al.,** Ordered Sets and Lattices II
151 **I. A. Aleksandrov, L. A. Bokut′, and Yu. G. Reshetnyak, Editors,** Second Siberian Winter School "Algebra and Analysis"
150 **S. G. Gindikin, Editor,** Spectral Theory of Operators
149 **V. S. Afraĭmovich et al.,** Thirteen Papers in Algebra, Functional Analysis, Topology, and Probability, Translated from the Russian
148 **A. D. Aleksandrov, O. V. Belegradek, L. A. Bokut′, and Yu. L. Ershov, Editors,** First Siberian Winter School "Algebra and Analysis"
147 **I. G. Bashmakova et al.,** Nine Papers from the International Congress of Mathematicians, 1986
146 **L. A. Aĭzenberg et al.,** Fifteen Papers in Complex Analysis
145 **S. G. Dalalyan et al.,** Eight Papers Translated from the Russian
144 **S. D. Berman et al.,** Thirteen Papers Translated from the Russian
143 **V. A. Belonogov et al.,** Eight Papers Translated from the Russian
142 **M. B. Abalovich et al.,** Ten Papers Translated from the Russian
141 **H. Draškovičová et al.,** Ordered Sets and Lattices
140 **V. I. Bernik et al.,** Eleven Papers Translated from the Russian
139 **A. Ya. Aĭzenshtat et al.,** Nineteen Papers on Algebraic Semigroups
138 **I. V. Kovalishina and V. P. Potapov,** Seven Papers Translated from the Russian
137 **V. I. Arnol′d et al.,** Fourteen Papers Translated from the Russian
136 **L. A. Aksent′ev et al.,** Fourteen Papers Translated from the Russian
135 **S. N. Artemov et al.,** Six Papers in Logic
134 **A. Ya. Aĭzenshtat et al.,** Fourteen Papers Translated from the Russian
133 **R. R. Suncheleev et al.,** Thirteen Papers in Analysis
132 **I. G. Dmitriev et al.,** Thirteen Papers in Algebra
131 **V. A. Zmorovich et al.,** Ten Papers in Analysis
130 **M. M. Lavrent′ev, K. G. Reznitskaya, and V. G. Yakhno,** One-dimensional Inverse Problems of Mathematical Physics
129 **S. Ya. Khavinson,** Two Papers on Extremal Problems in Complex Analysis
128 **I. K. Zhuk et al.,** Thirteen Papers in Algebra and Number Theory
127 **P. L. Shabalin et al.,** Eleven Papers in Analysis
126 **S. A. Akhmedov et al.,** Eleven Papers on Differential Equations
125 **D. V. Anosov et al.,** Seven Papers in Applied Mathematics
124 **B. P. Allakhverdiev et al.,** Fifteen Papers on Functional Analysis
123 **V. G. Maz′ya et al.,** Elliptic Boundary Value Problems
122 **N. U. Arakelyan et al.,** Ten Papers on Complex Analysis

(Continued in the back of this publication)

Selected Papers on Number Theory, Algebraic Geometry, and Differential Geometry

American Mathematical Society

TRANSLATIONS

Series 2 • Volume 160

Selected Papers on
Number Theory,
Algebraic Geometry,
and Differential Geometry

Katsumi Nomizu
Editor

American Mathematical Society
Providence, Rhode Island

Translation edited by KATSUMI NOMIZU

1991 *Mathematics Subject Classification.* Primary
11G40, 14C30, 14J27, 20D08, 53C55, 58G25

Library of Congress Cataloging-in-Publication Data
Selected papers on number theory, algebraic geometry, and differential geometry/Katsumi Nomizu, editor.   p. cm. — (American Mathematical Society translations, ISSN 0065-9290; ser. 2, v. 160)
Includes bibliographical references.
ISBN 0-8218-7511-6
1. Number theory. 2. Geometry, Algebraic. 3. Geometry, Differential. I. Nomizu, Katsumi, 1924–. II. Series.
QA3.A572   ser. 2, vol. 160
[QA241]
510 s—dc20   94-26691
[516.3′5]   CIP

Copying and reprinting. Individual readers of this publication, and nonprofit libraries acting for them, are permitted to make fair use of the material, such as to copy an article for use in teaching or research. Permission is granted to quote brief passages from this publication in reviews, provided the customary acknowledgment of the source is given.

Republication, systematic copying, or multiple reproduction of any material in this publication (including abstracts) is permitted only under license from the American Mathematical Society. Requests for such permission should be addressed to the Manager of Editorial Services, American Mathematical Society, P.O. Box 6248, Providence, Rhode Island 02940-6248. Requests can also be made by e-mail to reprint-permission@math.ams.org.

The appearance of the code on the first page of an article in this publication (including abstracts) indicates the copyright owner's consent for copying beyond that permitted by Sections 107 or 108 of the U.S. Copyright Law, provided that the fee of $1.00 plus $.25 per page for each copy be paid directly to the Copyright Clearance Center, Inc., 222 Rosewood Drive, Danvers, Massachusetts 01923. This consent does not extend to other kinds of copying, such as copying for general distribution, for advertising or promotional purposes, for creating new collective works, or for resale.

© Copyright 1994 by the American Mathematical Society. All rights reserved.
The American Mathematical Society retains all rights
except those granted to the United States Government.
Printed in the United States of America.

The paper used in this book is acid-free and falls within the guidelines
established to ensure permanence and durability.
Printed on recycled paper.
This volume was typeset using $\mathcal{A}_{\mathcal{M}}\mathcal{S}$-TEX,
the American Mathematical Society's TEX macro system.

10 9 8 7 6 5 4 3 2 1   99 98 97 96 95 94

Contents

Preface	ix
On the Number-Theoretic Method in Geometry: Geometric Analogue of Zeta and L-Functions and its Applications TOSHIKAZU SUNADA	1
Fundamental Groups and Laplacians TOSHIKAZU SUNADA	19
Moonshine: A Mysterious Relationship between Simple Groups and Automorphic Functions MASAO KOIKE	33
On the Theory of Mixed Hodge Modules MORIHIKO SAITO	47
Integral Invariants in Kähler Geometry AKITO FUTAKI	63
A Convergence Theorem for Einstein Metrics and the ALE Spaces HIRAKU NAKAJIMA	79
On the Topology of Elliptic Surfaces—a Survey MASAAKI UE	95
Modular p-adic L-functions and p-adic Hecke Algebras HARUZO HIDA	125

Preface

This is a collection of several papers that originally appeared in the journal Sugaku in Japanese. These translated articles would normally appear in the AMS journal **Sugaku Expositions**. In order to expedite publication, we have chosen, with the consent of the Mathematical Society of Japan, to put them together in this form of selected papers and publish it as a volume of the Society's Translations Series 2.

This volume contains papers on Number Theory, Algebraic Geometry, and Differential Geometry. The first two papers by Sunada establish a close relationship between number theory and differential geometry. The papers by Nakajima and Futaki are cross-referenced between algebraic and differential geometry. Each of the other four papers by Koike, Saito, Ue, and Hida deals with a subject where algebraic geometry and number theory come into closest contact.

On the Number-Theoretic Method in Geometry: Geometric Analogue of Zeta and L-Functions and its Applications

Toshikazu Sunada

§0. The Riemann zeta function

It is sometimes the case that a function which is defined, at first sight, in a simple way turns out to have unexpected rich properties and to have structures that can be generalized in various directions. A typical example is the Riemann zeta function. Indeed, the notion of a zeta function is generalized immediately to the case of arbitrary algebraic number fields (Dedekind zeta functions), and further to Artin L-functions involving representations of Galois groups and class field theory. The analogue of a Dedekind zeta function for a function field over a finite field, which is called a congruence zeta function, is a special case of the Weil zeta function of a general algebraic variety over a finite field. See the section on zeta functions of *Suugaku jiten*, Iwanami (English translation; *Encyclopedic Dictionary of Mathematics*, MIT press, 1980), for general properties of zeta and L-functions in number theory.

Let us recall the definition and some of the properties of the Riemann zeta function. For a complex number s, the series $\sum_{n=1}^{\infty} n^{-s}$ converges absolutely if $\operatorname{Re} s > 1$, and converges uniformly in the right half plane $\operatorname{Re} s \geq 1+\varepsilon$ (for any $\varepsilon > 0$). Hence, the function $\zeta(s) = \sum_{n=1}^{\infty} n^{-s}$ is a regular function on the right half plane $\operatorname{Re} s > 1$ and is called the Riemann zeta function.

PROPOSITION 1. (1) (Euler product representation) $\zeta(s) = \prod_p (1 - p^{-s})^{-1}$ if $\operatorname{Re} s > 1$, where p runs through all prime numbers.

(2) $\zeta(s)$ is meromorphically continued to all points in the s-plane.

(3) $\zeta(s)$ has a pole only at $s = 1$, and the pole is simple.

(4) $\zeta(s)$ has no zeros in the domain $\operatorname{Re} s \geq 1$.

(5) (The Riemann hypothesis; the validity is not known at present, December, 1993.)
The zeros of $\zeta(s)$ in the domain $0 < \operatorname{Re} s < 1$ are all on the line $\operatorname{Re} s = 1/2$.

(6) $\zeta(s)$ *satisfies a functional equation.*

It is a well-known fact in analytic number theory that the prime number theorem results from the above properties (1)–(4).

1991 *Mathematics Subject Classification.* Primary 58G25.
This article originally appeared in Japanese in Sūgaku **38** (4) (1986), 289–301.

In this exposition, we introduce some analogues in geometry (especially in spectral geometry) of zeta and L-functions and give their applications. Among other things, we would like to emphasize that some geometric objects (e.g., closed geodesics) have properties that are similar to (or the same as) number-theoretic objects, and to show that the consideration of the geometric analogue of zeta functions is by no means unnatural.

A remark before we go into our subject: what basis should we rely upon when we claim that a function is an analogue of a zeta or an L-function? Some candidates as key properties for this are: (a) it is a Dirichlet series; (b) it has an Euler product; (c) it is analytically continued to all points in the plane; (d) it satisfies a functional equation. The only common property of the functions we treat in this exposition is (a). Some functions satisfy (b) but not (c), and vice versa. Especially, (d) can hardly be expected except for some cases in which there exist certain special geometric structures.

§1. Zeta functions associated with elliptic differential operators

Let M be a compact connected C^∞-Riemann manifold, let $E \to M$ be a C^∞-vector bundle with a Hermitian metric, and let $A \colon C^\infty(E) \to C^\infty(E)$ be a formally selfadjoint elliptic differential operator acting on the sections of E, and assume further that A is nonnegative (i.e., $(As, s) \geq 0$ for all $s \in C^\infty(E)$). Then it is known that the spectrum of A consists only of nonnegative eigenvalues with finite multiplicity and, if $\{0 \leq \lambda_0 \leq \lambda_1 \leq \cdots\}$ is the set of eigenvalues with multiplicity, then the series

$$\zeta_A(s) = \sum_{\lambda_i > 0} \lambda_i^{-s}$$

converges absolutely for complex numbers s with $\operatorname{Re} s$ sufficiently large. We call $\zeta_A(s)$ the *zeta function* of A.

EXAMPLE. Given a lattice $\Gamma \subset \mathbb{R}^n$, we obtain a torus $\mathbb{R}^n/\Gamma$ $(= T_\Gamma)$ with the flat metric. The zeta function $\zeta_\Gamma(s)$ of the Laplacian Δ_Γ acting on the functions on T_Γ is equal to

$$(4\pi^2)^{-s} \sum_{\eta \in \Gamma^* - 0} \|\eta\|^{-2s}.$$

Here $\Gamma^* = \{\eta \in \mathbb{R}^n; \langle \eta, \gamma \rangle \in \mathbb{Z} \text{ for all } \gamma \in \Gamma\}$ denotes the dual lattice of Γ. Indeed, the eigenvalues and eigenfunctions are given, respectively, by $4\pi^2 \|\eta\|^2$ and $\exp 2\pi\sqrt{-1}\langle \lambda, \eta \rangle$ for various $\eta \in \Gamma^*$. Considering lattices in $\mathbb{R}^n$ is equivalent to considering bilinear forms on $\mathbb{R}^n$, and consequently, the Epstein zeta function of a bilinear form is essentially equal to our $\zeta_\Gamma(s)$. Hence, $\zeta_\Gamma(s)$ is meromorphically continued to the whole s-plane and satisfies a functional equation.

PROPOSITION 2. (Seeley [48]) *For any nonnegative self-adjoint elliptic differential operator A, we have*

(1) $\zeta_A(s)$ *is meromorphically continued to the whole s-plane and is regular at $s = 0$.*

(2) $\zeta_A(s)$ *can be, in principle, written in terms of the "geometric data" of A, E, and M.*

APPLICATION. For any elliptic differential operator $P \colon C^\infty(E) \to C^\infty(F)$, we define the index $\operatorname{Ind} P$ of P to be $\dim(\operatorname{Ker} P) - \dim(\operatorname{Coker} P)$. From (2) above,

Ind P is seen to be written in terms of the "geometric data" of P, E, F, and M as follows:

Let P^* denote the formal adjoint of P, and set
$$C_\lambda(E) = \{u \in C^\infty(E); P^*Pu = \lambda u\},$$
$$C_\lambda(F) = \{v \in C^\infty(F); PP^*v = \lambda v\}.$$

Then $P(C_\lambda(E)) \subset C_\lambda(F)$ and $P^*(C_\lambda(F)) \subset C_\lambda(E)$. If $\lambda > 0$, then $P|C_\lambda(E)$ and $P^*|C_\lambda(F)$ are injective; so $\dim C_\lambda(E) = \dim C_\lambda(F)$. Hence, setting $A_1 = P^*P + I$ and $A_2 = PP^* + I$, we see that
$$\zeta_{A_1}(s) - \zeta_{A_2}(s) = \sum (\lambda + 1)^{-s} - \sum (\mu + 1)^{-s}$$
$$= \dim C_0(E) - \dim C_0(F) = \text{Ind } P.$$

Thus, we obtain $\text{Ind } P = \zeta_{A_1}(0) - \zeta_{A_2}(0)$, an analytic representation for $\text{Ind } P$.

For some special operators (e.g., the Dirac operator), one can apply geometric invariant theory to obtain an analytic proof of the Atiyah-Singer index theorem [19].

§2. Isospectral manifolds

Take, as our elliptic differential operator A, the Laplacian $\Delta_{p,M}$ on the space of p-forms ($0 \le p \le \dim M$), and set $\zeta_{p,M}(s) = \zeta_{\Delta_{p,M}}(s)$. The following fact can be easily shown.

PROPOSITION 3. *We have $\zeta_{p,M_1}(s) = \zeta_{p,M_2}(s)$ if and only if Δ_{p,M_1} and Δ_{p,M_2} have the same eigenvalues (including multiplicities).*

M_1 and M_2 are said to be isospectral if Δ_{p,M_1} and Δ_{p,M_2} have the same eigenvalues for all p.

It is an important subject of the spectral problem to study the relationship between two isospectral manifolds M_1 and M_2, and much has already been studied. Firstly, one has $\dim H_p(M_1, \mathbb{R}) = \dim H_p(M_2, \mathbb{R})$, because $\dim H_p(M, \mathbb{R}) =$ the multiplicity of the eigenvalue 0 of $\Delta_{p,M}$. However, J. Milnor [36] constructed an example of flat tori (of dimension 16) that are isospectral but are not isometric; A. Ikeda [23] constructed an example of manifolds in a family of spherical space forms that are isospectral but have different homotopy types; M. F. Vignéras [66] constructed an example of manifolds in a family of hyperbolic space forms that are isospectral but have nonisomorphic fundamental groups, and further in the case of dimension 2, she constructed surfaces with constant curvature that are isospectral but are not isometric. In the cases above, they exploited the specialty of the space forms and certain precise identities such as the Jacobi identity or its nonabelian version, the Selberg trace formula. Here, employing ideas from number theory, we show in a completely elementary manner the following fact.

THEOREM A. *There exist four-dimensional Riemannian manifolds M_1 and M_2 that are isospectral but $H_1(M_1, \mathbb{Z}) \not\cong H_1(M_2, \mathbb{Z})$.*

The key to the construction is that $\zeta_M(s)$ has the "same" property as the Dedekind zeta function of an algebraic number field. For example, we have the geometric analogue of the following proposition in number theory.

PROPOSITION 4. *Let K be a finite Galois extension of the rational number field $\mathbb{Q}$ with Galois group G. Let k_1 and k_2 be two subfields of K, and let H_1 and H_2*

be, respectively, the corresponding subgroups of G. Then the following conditions are equivalent:

(i) *For any conjugacy class of G, the intersections of the class with H_1 and H_2 have the same cardinality.*

(ii) *For any prime number p, either p ramifies in both k_1 and k_2, or p does not ramify and has the same type of decomposition in k_1 and k_2.*

(iii) *The zeta functions of k_1 and k_2 coincide.*

This proposition provides us with a method for constructing number fields k_1 and k_2 that have the same zeta function but are not isomorphic to each other. Indeed, if we find subgroups H_1 and H_2 that are not conjugate in G and satisfy condition (i), then the fields k_1 and k_2 give an example of such a couple (see, e.g., Komatsu [29]).

Now the geometric analogue of the notion of an extension field is, of course, a covering space. In the category of Riemannian manifolds, a Riemannian covering is a natural object, where we take into account the condition of local isometry.

The above Proposition is then translated exactly into the following Proposition.

PROPOSITION 5. *Let $\varpi\colon M \to M_0$ be a finite Galois Riemannian covering with transformation group G. Let $\varpi_1\colon M_1 \to M_0$ and $\varpi_2\colon M_2 \to M_0$ be the coverings corresponding to two subgroups H_1 and H_2 of G, respectively. If the triple (G, H_1, H_2) satisfies condition (i) of Proposition 4, then we have $\zeta_{p,M_1}(s) \equiv \zeta_{p,M_2}(s)$ for all p. In particular, M_1 and M_2 are isospectral. (However, even if H_1 and H_2 are not conjugate in G, M_1 and M_2 can be isometric.)*

Before going into the proof of Proposition 5, we first prove Theorem A. Take the following as our triple (G, H_1, H_2) (K. Komatsu [29]): let $p > 2$ be a prime number, and set

$$H_1 = (\mathbb{Z}/p\mathbb{Z})^3 \quad \text{(direct product)}$$

$$H_2 = (\mathbb{Z}/p\mathbb{Z})^3 \quad \text{with group structure defined by the rule}:$$

$$(k, m, n)(k', m', n') = (k + k' - nm', m + m', n + n').$$

H_2 is noncommutative, and we have

$$(k, m, n)^p = \left(pk - \frac{p(p-1)}{2}mn, pm, pn\right) = (0, 0, 0),$$

so that the exponent of H_2 is p. Identify H_1 and H_2 as sets, and write S for this set. Then each of H_1 and H_2 acts by left transformation on S and can be regarded as a subgroup of the permutation group $G = \mathfrak{S}(S)$. It is easily checked that this triple (G, H_1, H_2) satisfies (i) of Proposition 4.

Next we take, as our compact C^∞-manifold M_0, a manifold whose fundamental group $\pi_1(M_0)$ is isomorphic to G. This is actually possible, because it is known that any group with finite presentation can be realized as the fundamental group of a four-dimensional compact C^∞-manifold.

Let $\varpi_0\colon M \to M_0$ be the universal covering of M_0. If M_1 and M_2 are coverings of M_0 such that $\pi_1(M_1) = H_1$ and $\pi_1(M_2) = H_2$, then we have a commutative diagram (Figure 1) of covering maps. If we lift an *arbitrary* Riemannian metric g_0 on M_0 by the covering maps, then $\varpi_0, \varpi_1,$ and ϖ_2 become Riemannian coverings and, by

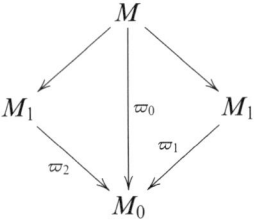

FIGURE 1

Proposition 5, $(M_1 \varpi_1^* g_0)$ and $(M_2, \varpi_2^* g_0)$ are isospectral. On the other hand, we have
$$H_1(M_1, \mathbb{Z}) = \pi_1(M_1) \neq H_1(M_2, \mathbb{Z}).$$

§3. An elementary trace formula

Let us now prove Proposition 5. The key to the proof is an elementary trace formula (Sunada [63]), which might be regarded as the prototype of the Selberg trace formula. Let V be a (separable) Hilbert space on which a finite group G acts unitarily. Let $A: V \to V$ be a nonnegative selfadjoint bounded operator which we assume belongs to the trace class and is compatible with the G-action.

LEMMA 1 (Elementary trace formula). *The restriction $A|V^G$ of A to the subspace $V^G = \{v \in V; gv = v \text{ for all } g \in G\}$ belongs to the trace class, and one has*
$$\mathrm{tr}(A|V^G) = \sum_{[g] \in [G]} (^\# G_g)^{-1} \mathrm{tr}(gA).$$

Here $[G] = \{[g]\}$ is the set of all conjugacy classes of G, and G_g is the centralizer of g.

PROOF. The orthogonal projection $P: V \to V^G$ is given by
$$P(v) = (^\# G)^{-1} \sum_{g \in G} g \cdot v.$$

Since $\mathrm{tr}(A|V^G) = \mathrm{tr}(PA)$, we get
$$\mathrm{tr}(A|V^G) = \sum_{g \in G} (^\# G)^{-1} \mathrm{tr}(gA).$$

Here we replace the sum $\sum_{g \in G}$ by the double sum $\sum_{[g] \in [G]} \sum_{a \in [g]}$, and we make the identification $G/G_g \leftrightarrow [g]$ ($a \leftrightarrow aga^{-1}$) to see
$$\sum_{a \in [g]} \mathrm{tr}(aA) = \sum_{a \in G/G_g} \mathrm{tr}(aga^{-1}A) = \sum_{a \in G/G_g} \mathrm{tr}(gA)$$
$$= {}^\#(G/G_g) \mathrm{tr}(gA).$$

Thus, we obtain the required trace formula. □

LEMMA 2. *If the triple (G, H_1, H_2) satisfies* (i) *of Proposition 4, then we have*
$$\operatorname{tr}(A|V^{H_1}) = \operatorname{tr}(A|V^{H_2}).$$

PROOF. It is enough to see
$$\operatorname{tr}(A|V^H) = (^{\#}H)^{-1} \sum_{[g]\in[G]} {}^{\#}([g]\cap H)\operatorname{tr}(gA)$$

for any subgroup H of G. Letting $[h]_H$ denote the conjugacy class in H of an element $h \in H$ and noticing that
$$\sum_{[h]_H \subset [g]\cap H} {}^{\#}[h]_H = {}^{\#}([g]\cap H)$$
$$[h]_H \subset [g]\cap H \Rightarrow \operatorname{tr}(hA) = \operatorname{tr}(gA),$$

we obtain the desired formula:
$$\operatorname{tr}(A|V^H) = (^{\#}H)^{-1} \sum_{[g]\in[G]} \sum_{[h]_H \subset [g]\cap H} {}^{\#}[h]_H \operatorname{tr}(hA)$$
$$= (^{\#}H)^{-1} \sum_{[g]\in[G]} {}^{\#}([g]\cap H)\operatorname{tr}(gA). \qquad \square$$

PROOF OF PROPOSITION 5. We only treat the case $p = 0$, i.e., the case of the Laplacian acting on the space of functions (the general case can be proved similarly). Apply Lemma 2 to the case
$$V = V_M = \left\{ f \in L^2(M); \int_M f\, dx = 0 \right\},$$
$$A = A_M = (\Delta_M|V)^{-s}, \qquad s \gg 0.$$

It is known that A_M belongs to the trace class and that $\operatorname{tr} A_M = \zeta_M(s)$. If $\varpi_i \colon M \to M_i$ denotes the covering map, then the linear map $L^2(M_i) \to L^2(M); f \mapsto (^{\#}H_i)^{-1/2} f \circ \varpi_i$ induces an isometry $\phi_i \colon V_{M_i} \cong V_M^{H_i}$ of Hilbert spaces, and $\phi_i \circ A_{M_i} = A_M \circ \phi_i$. Hence,
$$\zeta_{M_1}(s) = \operatorname{tr}(A_M|V_M^{H_1}) = \operatorname{tr}(A_M|V_M^{H_2}) = \zeta_{M_2}(s). \qquad \square$$

§4. A problem of Gel'fand

I. M. Gel'fand [18] raised the question whether a cocompact discrete subgroup Γ of $\operatorname{PSL}_2(\mathbb{R})$ is determined up to conjugacy by the induced representation of $\operatorname{PSL}_2(\mathbb{R})$ on $L^2(\Gamma\backslash\operatorname{PSL}_2(\mathbb{R}))$. This question is closely related to the isospectral problem of compact Riemann surfaces, and in Vignéras' paper [66] cited before, a counterexample was constructed within the family of arithmetic discrete subgroups (obtained from quaternion algebras over algebraic number fields). Such examples are necessarily found only sporadically, because it is known that there are only finitely many conjugacy classes of arithmetic discrete subgroups with a given signature. In this section, we will see that the idea of the last section can be used to construct a large number of counterexamples to the problem of Gel'fand (Sunada [60]).

The next proposition is deduced immediately from fundamental facts on induced representations.

PROPOSITION 6. *Let $\mathscr{G}$ be a unimodular locally compact group. Let $\Gamma, \Gamma_0, \Gamma_1$, and Γ_2 be cocompact discrete subgroups of $\mathscr{G}$ such that $\Gamma \subset \Gamma_i \subset \Gamma_0$ for $i = 1, 2$, and assume Γ is normal of finite index in Γ_0. If the triple $(\Gamma_0/\Gamma, \Gamma_1/\Gamma, \Gamma_2/\Gamma)$ satisfies (i) of Proposition 4, then the representations of $\mathscr{G}$ on $L^2(\Gamma_i \backslash \mathscr{G})$, $i = 1, 2$ are unitarily equivalent.*

Take as our group $\Gamma_0 \subset \mathrm{PSL}_2(\mathbb{R})$, a torsion-free subgroup that satisfies the following three conditions:

(a) The Riemann surface $\Gamma_0 \backslash \mathrm{PSL}_2(\mathbb{R})/\mathrm{SO}(2)$ has genus equal to or greater than 3.

(b) Γ_0 is maximal in $\mathrm{PGL}_2(\mathbb{R})$ as a discrete subgroup.

(c) Γ_0 is nonarithmetic.

The theory of Teichmüller spaces assures the existence of a Γ_0 that satisfies these conditions (in fact, a "generic" Γ_0 satisfies (a), (b), and (c)). Next we take as a triple (G, H_1, H_2) of finite groups satisfying (i) of Proposition 4, the following:

$$G = (\mathbb{Z}/8\mathbb{Z})^\times \cdot \mathbb{Z}/8\mathbb{Z} \quad \text{(semidirect product)},$$
$$H_1 = \{(1,0), (3,0), (5,0), (7,0)\},$$
$$H_2 = \{(1,0), (3,4), (5,4), (7,0)\}.$$

(Note that H_1 and H_2 are not conjugate in G and that G is generated by three elements.) On the other hand, if k denotes the genus of the Riemann surface $\Gamma_0 \backslash \mathrm{PSL}_2(\mathbb{R})/\mathrm{SO}(2)$, there is a surjective homomorphism of Γ_0 onto a free group generated by k elements; so we eventually get a surjective homomorphism $\varphi \colon \Gamma_0 \to G$. Let $\Gamma_i = \varphi^{-1}(H_i)$ for $i = 1, 2$. By the above Proposition, Γ_1 and Γ_2 yield unitarily equivalent representations of $\mathrm{PSL}_2(\mathbb{R})$. Thus, the only assertion that remains to be proved is that Γ_1 and Γ_2 are not conjugate in $\mathrm{PGL}_2(\mathbb{R})$. If we assume they are conjugate, then $g\Gamma_1 g^{-1} = \Gamma_2$ for some $g \in \mathrm{PGL}_2(\mathbb{R})$, whereas $g\Gamma_0 g^{-1}$ and Γ_0 are commensurable. According to Margulis, however, for a nonarithmetic Γ_0, the set

$$C(\Gamma_0) = \{h \in \mathrm{PGL}_2(\mathbb{R}) \colon h\Gamma_0 h^{-1} \text{ and } \Gamma_0 \text{ are commensurable}\}$$

forms a discrete subgroup of $\mathrm{PGL}_2(\mathbb{R})$ that contains Γ_0. The maximality of Γ_0 implies $C(\Gamma_0) = \Gamma_0$. We thus have $g \in \Gamma_0$, which leads to a contradiction $\varphi(g) H_1 \varphi(g)^{-1} = H_2$.

REMARK. For Γ_i as above, the genus of the Riemann surface $\Gamma_i \backslash \mathrm{PSL}_2(\mathbb{R})/\mathrm{SO}(2)$ is $8k - 7$ (if $k \geq 3$). Recently, Buser [13] and Brooks-Tse [12] constructed examples of genus ≥ 4 by methods similar to those above.

§5. "Class field theory" for geodesics

Let M be a compact connected Riemannian manifold of dimension ≥ 2. A closed geodesic on M is a smooth closed curve on M whose length is locally minimal; hereafter we think of a closed geodesic as an oriented cycle, regardless of parametrization. Further, we call a closed geodesic *prime* if it is not a multiple of another geodesic.

Let $\varpi \colon M \to M_0$ be an n-fold covering, and let $\mathfrak{P}$ be a prime closed geodesic on M. The image $\varpi(\mathfrak{P})$ is not prime in general. If $\varpi(\mathfrak{P})$ is an m-multiple of a prime closed geodesic $\mathfrak{p}$ on M_0, $\varpi(\mathfrak{P}) = \mathfrak{p}^m$, then we call m the *degree* of $\mathfrak{P}$ and say that

$\mathfrak{P}$ is a lift of $\mathfrak{p}$. There are only finitely many lifts of $\mathfrak{p}$. If $\mathfrak{P}_1, \ldots, \mathfrak{P}_g$ are all of them, then we have

$$\sum_{i=1}^{g} \text{degree } \mathfrak{P}_i = n.$$

This shows that the relation between finite extensions of algebraic number fields and prime ideal decomposition holds also between finite coverings and prime closed geodesics in exactly the same form. If $\varpi: M \to M_0$ is furthermore a Galois covering, then the covering transformation group G acts on the set $\{\mathfrak{P}_1, \ldots, \mathfrak{P}_g\}$ transitively and, for each lift $\mathfrak{P}$ of $\mathfrak{p}$, there exists an element $(\mathfrak{P}|\varpi)$ of G that has the following properties:

$$(\mathfrak{P}|\varpi)\mathfrak{P} = \mathfrak{P},$$
$$(g\mathfrak{P}|\varpi) = g(\mathfrak{P}|\varpi)g^{-1} \quad \text{for all } g \in G,$$
$$\{g \in G; g\mathfrak{P} = \mathfrak{P}\} = \langle(\mathfrak{P}|\varpi)\rangle.$$

$(\mathfrak{P}|\varpi)$ is the analogue of the Frobenius substitution of number fields. When the covering ϖ is abelian, i.e., G is abelian, we may write $(\mathfrak{P}|\varpi) = (\mathfrak{p}|\varpi)$ because $(\mathfrak{P}|\varpi)$ depends only on the prime $\mathfrak{p}$ lying in the image of $\mathfrak{P}$.

Next let us consider a free abelian group I_M with free basis consisting of prime geodesics on M. This replaces the ideal group in number theory (by a theorem of Dedekind, the multiplicative group of fractional ideals of a number field is a free abelian group with free basis the prime ideals). If we denote multiplicatively, then I_M is the set of all formal products

$$\mathfrak{a} = \mathfrak{p}_1^{a_1} \cdots \mathfrak{p}_k^{a_k}, \qquad a_i \in \mathbb{Z}.$$

We regard $\mathfrak{a}$ as a one-dimensional cycle on M and call $\mathfrak{a}$ *principal* if it is homologous to zero. Accordingly, we take I_M/I_M^0 as the analogue of the (absolute) ideal class group. Here I_M^0 is the subgroup of all principal elements $\mathfrak{a} \in I_M$. Noticing that every homology class contains a closed geodesic, we see $I_M/I_M^0 = H_1(M, \mathbb{Z})$.

Given a covering $\varpi: M \to M_0$, we define a homomorphism $N: I_M \to I_{M_0}$ by $N(\mathfrak{P}) = \mathfrak{p}^{\text{degree } \mathfrak{P}} = \varpi(\mathfrak{P})$.

The following proposition is a geometric analogue of the fundamental theorem of class field theory.

PROPOSITION 7. (i) *We have* $[I_{M_0} : I_{M_0}^0 \cdot N(I_M)] \leq n$. *The equality holds if and only if the covering is abelian.*

(ii) *If ϖ is abelian, the correspondence $\mathfrak{p} \mapsto (\mathfrak{p}|\varpi)$ induces an isomorphism $I_{M_0}/I_{M_0}^0 \cdot N(I_M) \to G$.*

(iii) *For any subgroup H of I_{M_0} that is of finite index and contains $I_{M_0}^0$, there exists an abelian covering $\varpi: M \to M_0$ such that $H = I_{M_0}^0 \cdot N(I_M)$.*

The proof is almost trivial, compared with the one for the original class field theory for algebraic number fields.

First, the diagram of Figure 2 is commutative by the definition of the homomorphism $N: I_M \to I_{M_0}$. So $I_{M_0}/I_{M_0}^0 \cdot N(I_M)$ is isomorphic to $H_1(M_0, \mathbb{Z})/\varpi_*(H_1(M, \mathbb{Z}))$. On the other hand, a surjective map

$$\Phi: \pi_1(M_0)/\varpi_*(\pi_1(M)) \to H_1(M_0, \mathbb{Z})/\varpi_*(H_1(M, \mathbb{Z}))$$

is defined according to the commutative diagram in Figure 3. This together with the

$$I_M \xrightarrow{N} I_{M_0}$$
$$\downarrow \qquad \qquad \downarrow$$
$$H_1(M, Z) \xrightarrow{\varpi_*} H_1(M_0, \mathbb{Z})$$

FIGURE 2

$$\pi_1(M) \xrightarrow{\varpi_*} \pi_1(M_0)$$
$$\downarrow \qquad \qquad \downarrow$$
$$H_1(M, \mathbb{Z}) \xrightarrow{\varpi_*} H_1(M_0, \mathbb{Z})$$

FIGURE 3

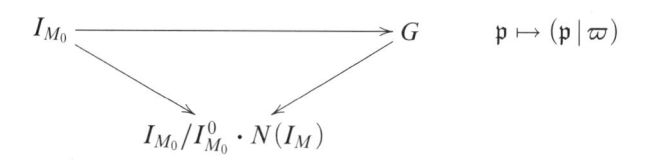

$$I_{M_0} \longrightarrow G \qquad \mathfrak{p} \mapsto (\mathfrak{p} \mid \varpi)$$
$$\searrow \qquad \nearrow$$
$$I_{M_0}/I_{M_0}^0 \cdot N(I_M)$$

FIGURE 4

fact that $^\#(\pi_1(M_0)/\varpi_*(\pi_1(M))) = n$ implies the required inequality of (i). If the equality holds, then Φ is bijective; so $\varpi_*(\pi_1(M))$ is normal in $\pi_1(M_0)$ and Φ is a group homomorphism.

To show (ii), it suffices to show the commutativity of the diagram in Figure 4. This is clear from the definition of $(\mathfrak{p}|\varpi)$.

Let us check (iii). There is a one-to-one correspondence between any two of the following three sets: the set of subgroups of I_{M_0} that are of finite index and contain $I_{M_0}^0$, the set of subgroups H' of $H_1(M_0, \mathbb{Z})$ that are of finite index, and the set of subgroups Γ of $\pi_1(M_0)$ that are of finite index and contain the commutator subgroup $[\pi_1(M_0), \pi_1(M_0)]$. If Γ corresponds to H, then $H = I_{M_0}^0 \cdot N(I_M)$ for the covering $\varpi \colon M = \widetilde{M_0}/\Gamma \to M_0$.

Next let us see that there exists, also in the theory of geodesics, an analogue of Proposition 4.

PROPOSITION 8. *Let the notation and assumptions be as in Proposition 5. Then the triple* (G, H_1, H_2) *satisfies condition* (i) *of Proposition 4 if and only if any prime closed geodesic* $\mathfrak{p}$ *on* M_0 *decomposes in the same manner in* M_1 *and in* M_2, *i.e., letting* $\mathfrak{P}_1, \ldots, \mathfrak{P}_g$ *and* $\mathfrak{P}'_1, \ldots, \mathfrak{P}'_{g'}$ *be the lifts of* $\mathfrak{p}$ *to* M_1 *and* M_2, *respectively, we have* $g = g'$ *and* $\deg \mathfrak{P}_i = \deg \mathfrak{P}'_i$ *for* $i = 1, \ldots, g$ (*after a suitable renumbering*).

PROOF. The triple (G, H_1, H_2) satisfies condition (i) of Proposition 4 if and only if the permutation representations of G on G/H_1 and G/H_2 are equivalent as linear representations. Recall that a linear representation of a finite group is determined completely by its character and that the character of the permutation representation of G with respect to a subgroup H is given by

$$\chi(g) = {}^\#\{g_1 H \in G/H; g_1^{-1} g g_1 \in H\}$$
$$= {}^\#G_g \cdot {}^\#([g] \cap H)/{}^\#H$$

for $g \in G$. Let $M_H \to M_0$ be the covering map corresponding to H. Let $\mathfrak{P}$ be a lift to M of a prime closed geodesic $\mathfrak{p}$ on M_0, and let $\{\mathfrak{q}_1, \ldots, \mathfrak{q}_r\}$ be the set of lifts of $\mathfrak{p}$ to M_H. If we choose $\tau_1, \ldots, \tau_r \in G$ such that $\tau_i \mathfrak{P}$ is a lift of $\mathfrak{q}_i$ for each i, then

$$G = \bigcup_{i=1}^{r} \bigcup_{x_i=0}^{f_i-1} H\tau_i \sigma^{-x_i} \quad \text{(disjoint)}.$$

Here $\sigma = (\mathfrak{P}|\varpi)$ and $f_i = \deg \mathfrak{q}_i$. By the above character formula, we have for any m,

$$\chi(\sigma^m) = {}^\#\{\tau_i \sigma^{x_i}; 1 \leq i \leq r, \, 0 \leq x_i \leq f_i - 1, \, \tau_i \sigma^{x_i} \sigma^m \sigma^{-x_i} \tau_i^{-1} \in H\}$$
$$= \sum_{\substack{1 \leq i \leq r \\ \tau_i \sigma^m \tau_i^{-1} \in H}} f_i = \sum_{\substack{1 \leq i \leq r \\ f_i | m}} f_i.$$

On the other hand, any element of G can be written in the form $(\mathfrak{P}|\varpi_0)^m$ for some $\mathfrak{P}$ and m, since any free homotopy class contains at least one closed geodesic. Hence, the decomposition law of prime closed geodesics for the covering $M_H \to M_0$ determines completely the character χ. The converse implication is of course valid. □

We write $l(\mathfrak{p})$ for the length of a closed geodesic $\mathfrak{p}$.

COROLLARY. *Under condition* (i) *of Proposition* 4, M_1 *and* M_2 *are isolength spectral in the sense that for each* $x \geq 0$, *there exists a one-to-one correspondence between the sets* {*prime closed geodesics* $\mathfrak{q}_1$ *on* M_1; $l(\mathfrak{q}_1) = x$} *and* {*prime closed geodesics* $\mathfrak{q}_2$ *on* M_2; $l(\mathfrak{q}_2) = x$}.

REMARK. It is known that, for a "generic" metric, the eigenvalue spectrum determines the length spectrum.

§6. Closed geodesics and L-functions

It is a well-known fact that there exist infinitely many prime ideals in an algebraic number field. Is there an analogue of this for Riemannian manifolds? That is, do there exist infinitely many prime closed geodesics on a compact Riemannian manifold? This is an important problem in differential geometry, but a complete proof has not yet been obtained (a "proof" of this is given in Klingenberg [28], but to the best of the authors' knowledge, the gap does not seem to have been filled completely).

But the affirmative answer to the above, together with more precise information, can be obtained for some special classes of manifolds. Before stating our result, let us refer to Dirichlet's prime number theorem for arithmetic progressions, which is a model for our argument.

PROPOSITION 9. *Let a and n be coprime natural numbers. Then there exist infinitely many primes in the arithmetic progression* $a, a+n, a+2n, \ldots$. *Moreover, denoting by $\pi(x; a, n)$ the number of primes not exceeding x that appear in this arithmetic progression, we have*

$$\lim_{x \to \infty} \frac{\pi(x; a, n)}{x/\log x} = \{{}^\#(\mathbb{Z}/n\mathbb{Z})^\times\}^{-1}.$$

Now the multiplicative group $(\mathbb{Z}/n\mathbb{Z})^\times$ is the Galois group of the cyclotomic field $\mathbb{Q}(e^{2\pi\sqrt{-1}/n})$ over $\mathbb{Q}$, and at the same time is the ideal class group (in the generalized sense) of that cyclotomic field. Thinking of $H_1(M,\mathbb{Z})$, as in §4, as the analogue of the absolute ideal class group, one may expect that

"Each homology class in $H_1(M,\mathbb{Z})$ contains infinitely many prime closed geodesics."

But one immediately finds a counterexample to this too optimistic conjecture. For example, if $M = \mathbb{R}^2/\mathbb{Z}^2$ is a flat torus, then $H_1(M,\mathbb{Z}) = \mathbb{Z}^2$, and $\alpha \in \mathbb{Z}^2$ contains a prime closed geodesic if and only if α is primitive, i.e., it is not a multiple of another homology class. On the other hand, if M is a closed surface of genus ≥ 2, then a combinatorial and group theoretic consideration on $\pi_1(M)$ shows that the above conjecture is true.

Now note that a closed surface of genus ≥ 2 can be endowed with a Riemannian metric of negative constant curvature. So it would be natural to expect the validity of the above conjecture for a general negatively curved manifold. Indeed, we have

THEOREM B. *Let M be a compact negatively curved Riemannian manifold. For each $\alpha \in H_1(M,\mathbb{Z})$ set $\pi(x,\alpha) = {}^{\#}\{\mathfrak{p} : $ prime closed geodesics on $M; l(\mathfrak{p}) < x, [\mathfrak{p}] = \alpha\}$ ([∘] denotes the homology class of ∘).*

(i) (*Adichi-Sunada* [6]).

$$\lim_{x \to \infty} \frac{1}{x} \log \pi(x,\alpha) = h > 0.$$

Here h is equal to the exponential degree of increase of the geodesic ball of the universal covering of M with respect to the radius

$$\lim_{R \to \infty} \frac{1}{R} \log \mathrm{vol}(B_R(\tilde{x}_0))$$

($B_R(\tilde{x}_0) = \{\tilde{x} \in \widetilde{M}; d(\tilde{x},\tilde{x}_0) \leq R\}$). In particular, there exist infinitely many prime closed geodesics in each α.

(ii) (Parry-Pollicott [41], Adachi-Sunada [5]). *If $H_1(M,\mathbb{Z})$ is of finite order, then*

$$\lim_{x \to \infty} \frac{\pi(x,\alpha)}{e^{hx}/hx} = \{{}^{\#}H_1(M,\mathbb{Z})\}^{-1}.$$

REMARK. (1) There exist compact negatively curved manifolds M (of dimension ≥ 3) with arbitrarily large $\dim H_1(M,\mathbb{R})$ (Milson [35]). $H_1(M,\mathbb{Z})$ is always finite when M is a Hamilton or Cayley hyperbolic space form.

(2) The above assertion holds true more generally, when the geodesic flow of M is of Anosov type. Further, for nonpositively curved manifolds M of rank one, Katsuda [25] showed that

$$\liminf_{x \to \infty} \frac{1}{x} \log \pi(x,\alpha) \geq \frac{1}{2}h.$$

(3) Let (X,φ_t) be a smooth Anosov flow on a compact manifold X, and set $\pi(x) = {}^{\#}\{\mathfrak{p}: $ periodic orbit of (φ_t) whose period $< x\}$. Then it is known that

$$\lim_{x \to \infty} \frac{1}{x} \log \pi(x) = h = \text{topological entropy of } (X,\varphi_t)$$

(and that $\lim_{x \to \infty} \pi(x)hx/e^{hx} = 1$ if (X,φ_t) has a topological mixing property). But for a general Anosov flow, a homology class in $H_1(X,\mathbb{Z})$ may happen to contain only a finite number of periodic orbits.

Here we shall give an outline of the proof of (ii), in which a number-theoretic idea can be clearly seen. An intricate combinatorial argument is needed to show (i).

First recall the method from analytic number theory used to prove Dirichlet's density theorem. Given a character χ of $(\mathbb{Z}/n\mathbb{Z})^\times$ of degree one, we define the L-function associated with χ by

$$L(s,\chi) = \prod_p (1 - \chi(p)p^{-s})^{-1}.$$

The following properties of $L(s,\chi)$ are known:
(1) $L(s,\chi)$ converges absolutely and is regular in the right half-plane $\operatorname{Re} s > 1$.
(2) $L(s,\chi)$ is continued analytically to the whole s-plane.
(3) $L(s,\chi)$ does not have a zero in the right half-plane $\operatorname{Re} s \geq 1$.
(4) $L(s,\chi)$ is regular in the whole s-plane if $\chi \neq 1$ (the trivial character).
(5) $L(s,1) = \zeta(s)$.

Hence, the Dirichlet series

$$-\sum_\chi \chi(-a)\frac{L'(s,\chi)}{L(s,\chi)} = \left(\sum_{\substack{p,k \\ p^k \equiv a \pmod n}} (\log p)p^{-ks}\right) \times {}^\#(\mathbb{Z}/n\mathbb{Z})^\times$$

has only a simple pole at $s = 1$ with residue 1. From this one obtains Proposition 9 by applying a theorem of Tauber type.

Now what is the geometric analogue of an L-function? Here we consider a function defined as follows. Given a one-dimensional unitary representation χ of $H_1(M,\mathbb{Z})$ (we do not assume the finiteness of $H_1(M,\mathbb{Z})$), we set

$$L(s,\chi) = \prod_{\mathfrak{p}}(1 - \chi([\mathfrak{p}])e^{-sl(\mathfrak{p})})^{-1}.$$

We expect this L-function to have similar properties to number-theoretic L-functions. For some special cases, the following is already known by Selberg [49]:

PROPOSITION 10. *Let M be a compact surface with a negative constant curvature ($\equiv 1$). Then the function (the Selberg zeta function)*

$$Z(s,\chi) = \prod_{k=0}^\infty \prod_{\mathfrak{p}}(1 - \chi[\mathfrak{p}])e^{-(s+k)l(\mathfrak{p})})$$

can be continued analytically to the whole s-plane. It has a simple zero at $s = 1$ if $\chi = 1$. For a general χ, its only zeros in the domain $\operatorname{Re} s \geq 0$ are

$$s = \frac{1}{2}\left(1 \pm \sqrt{1 - 4\lambda_k(\chi)}\right).$$

Here $\{\lambda_k(\chi)\}$ is the set of eigenvalues of the Laplacian on the flat line bundle on M corresponding to the representation χ. If $\chi \neq 1$ in particular, then $Z(s,\chi)$ is regular and has no zeros in the right half-plane $\operatorname{Re} s \geq 1$.

COROLLARY. $L(s,\chi) = Z(s+1,\chi)/Z(s,\chi)$ *has similar properties to number theoretic L-functions.*

To show the above proposition, Selberg made use of a trace formula which he invented. It is a formula describing a duality relation in the weak sense between the two sequences $\{l(\mathfrak{p})\}$ and $\{\lambda_k(\chi)\}$, and one cannot obtain such a precise trace formula for a general manifold with negative curvature. (One can obtain an asymptotic formula [55], but this is useless for our purpose.) Nevertheless, the following can be shown:

THEOREM C. (Parry-Pollicott [41] in case $\operatorname{Im}\chi$ is finite, Adachi-Sunada [5] in general).
(1) $L(s,\chi)$ converges absolutely and is regular in the right half-plane $\operatorname{Re} s > h$.
(2) $L(s,\chi)$ can be continued meromorphically to an open set D that contains the right half-plane $\operatorname{Re} s \geq h$.
(3) $L(s,\chi)$ has no zeros in D.
(4) $L(s,\chi)$ is regular in D if $\chi \neq 1$.
(5) $L(s,\chi)$ has a simple pole only at $s = h$ and is regular elsewhere.

REMARK. The above theorem is valid for the L-function of a more general Anosov flow (X, φ_t) and a finite-dimensional unitary representation of $\pi_1(X)$. See [5] for the details.

IDEA OF THE PROOF. For a general Anosov flow, the symbolic dynamics of Bowen [10] reduce the proof to considering the L-function of a pro-finite graph. Here we shall describe the L-function of a finite graph in order to clarify the outline of the proof.

Let (V, E) be a finite oriented graph; let V be the set of vertices, and let $E \subset V \times V$ be the set of edges. For each edge $e \in E$, denote by $o(e)$ and $t(e)$, respectively, the origin and terminal of e. Let l be a positive-valued function on E. A path of (V, E) is a sequence $c = (e_1, \ldots e_n)$ of edges $e_i \in E$ such that $t(e_i) = o(e_{i+1})$ for all $i = 1, \ldots, n-1$. We set $|c| = n$ and $l(c) = l(e_1) + \cdots + l(e_n)$. $c = (e_1, \ldots, e_n)$ is called a closed path if $t(e_n) = o(e_1)$. As in the case of closed geodesics, we consider the set of equivalence classes of closed paths neglecting parameterizations, and thus introduce the notions of a cycle and a prime cycle. (V, E) becomes in a natural way a one-dimensional CW-complex, whose fundamental group will be denoted by $\pi_1(V, E)$. For a unitary representation $\rho: \pi_1(V, E) \to U(N)$, we define its L-function $L(s, \rho)$ as follows:

$$L(s, \rho) = \prod_{\mathfrak{p}} \det(I_N - \rho(\langle \mathfrak{p} \rangle) \exp(-sl(\mathfrak{p})))^{-1}.$$

Here $\mathfrak{p}$ runs through the prime cycles of (V, E), and $\langle \mathfrak{p} \rangle$ is a representative in the conjugacy class of $\pi_1(V, E)$ corresponding to the free homotopy of $\mathfrak{p}$. $\square$

LEMMA 3. *If (V, E) is not homeomorphic to S^1, then there exists a positive number $h > 0$ such that*:
(1) $L(s, \rho)$ *converges absolutely in the right half-plane* $\operatorname{Re} s > h$;
(2) $L(s, \rho)$ *can be continued meromorphically to the whole s-plane*;
(3) $L(s, \rho)$ *has no zeros*;
(4) $L(s, 1)$ *has a simple pole at $s = h$*;
(5) *If $N \geq 2$ and ρ is irreducible, then $L(s, \rho)$ is regular in an open set that contains the right half-plane* $\operatorname{Re} s \geq h$.

PROOF. Let $(\widetilde{V}, \widetilde{E})$ be the universal covering of (V, E), so that $\pi_1(V, E)$ acts on $(\widetilde{V}, \widetilde{E})$ as an automorphism group. Define an endomorphism L_s of the linear space

Map($\widetilde{V}, \mathbb{C}^N$) by
$$L_s\varphi(x) = \sum_{\substack{e \in \widetilde{E} \\ o(e)=x}} \exp(-sl(e))\varphi(t(e)).$$

Define a finite-dimensional subspace F_ρ of Map($\widetilde{V}, \mathbb{C}^N$) by
$$F_\rho = \{\varphi; \varphi(\gamma x) = \rho(\gamma)\varphi(x) \text{ for all } \gamma \in \pi_1(V, E) \text{ and } x \in \widetilde{V}\}.$$

Then we have $L_s(F_\rho) \subset F_\rho$. After a simple calculation, we get
$$\mathrm{tr}(L_s|F_\rho)^k = \sum_{\substack{n, \mathfrak{p} \\ n|\mathfrak{p}|=k}} n^{-1} \mathrm{tr}\, \rho(\langle \mathfrak{p}\rangle^n) \exp(-snl(\mathfrak{p})).$$

So we have $L(s, \rho) = \det(I - L_s|F_\rho)^{-1}$, and the proof of the lemma is reduced to the study of eigenvalues of $L_s|F_\rho$. In the case $\rho = 1$, a classical theorem of Perron-Frobenius gives us the eigenvalues of $L_s|F_\rho$; in the general case, we complete the proof of the lemma by establishing the twisted version of the Perron-Frobenius theorem. □

§7. Complement

(1) Recently, L-functions have been considered in a highly general category by Kurokawa ([**30**]).

(2) Ihara's zeta functions for SL$_2$ over p-adic fields can be described in terms of finite graphs. They have meaning only for a special kind of nonoriented graph, whereas our L-functions are described in terms of general oriented graphs. (Ihara [**22**], Serre [**50**], Sunada [**62**]).

(3) Katsuda and the author (R. Phillips and P. Sarnak [**75**]) have recently obtained the following precise version of Theorem B, in the case where M is a surface with a negative constant curvature ($\equiv -1$) ([**68**]):
$$\pi(x, \alpha) \sim (g-1)^g \frac{e^x}{x^{g+1}} \qquad (x \uparrow \infty),$$
where g is the genus of M. A similar fact is conjectured to be true for a general manifold of negative curvature, but the proof would require a detailed consideration on the singularity of L-functions. Indeed, in the case of constant curvature, the above result is obtained by employing the Selberg zeta functions to study the variance of the poles of L-functions on the real line for various one-dimensional characters.

After the publication of this exposition, there was progress in several directions.

(1) A motivation for constructing isospectral manifolds can be found in the celebrated problem due to M. Kac, "Can one hear the shape of a drum?" This is the isospectral problem corresponding to the Dirichlet problem for two-dimensional domains, which had long been left unsolved. In 1991, however, C. Gordon, D. L. Webb, and S. Wolpert [**69**] extended the method stated in this exposition to the case of orbifolds to construct isospectral domains that are not congruent.

(2) C. Gordon and E. N. Wilson [**70**] constructed an example of isospectral deformations in the class of nilmanifolds, whereas D. M. DeTurck and C. S. Gordon [**71**] constructed, by generalizing the method of constructing isospectral manifolds explained in this exposition, such deformations using trace formulas.

(3) The isospectral potential problem for Schrödinger operators was considered by V. Gullemin and D. Kazhdan [**72**]; they asked if there exists an example of

Schrödinger operators on a compact manifold that have different potentials but have the same eigenvalues. In [72], it is proved that the eigenvalues determine the potential when the manifold is negatively curved and appropriately pinched. R. Brooks [73] used an interesting trick, in addition to the method in this exposition, to show that there exists an example of a negatively curved manifold on which the eigenvalues do not determine the potential.

(4) Employing dynamical L-functions, A. Katsuda and T. Sunada [74] proved, for general Anosov flows, Theorem B and the result stated in (3) above in a more precise form.

References

1. T. Adachi, *Markov families for Anosov flows with an involutive action*, Nagoya J. Math. **104** (1986), 55–62.
2. _____, *Closed orbits of Anosov flows and the fundamental group*, Proc. Amer. Math. Soc. **100** (1987), 595–598.
3. _____, *Distribution of closed geodesics with a preassigned homology class in a negatively curved manifold*, Nagoya Math. J. **110** (1988), 1–14.
4. T. Adachi and T. Sunada, *Energy spectrum of certain harmonic mappings*, Compositio Math. (1985).
5. _____, *Twisted Perron-Frobenius theorem and L-functions*, J. Funct. Anal. **71** (1986), 1–46.
6. _____, *Homology of closed geodesics in a negatively curved manifold*, J. Diff. Geom. **26** (1987), 81–99.
7. D. V. Anosov, *Ergodic properties of geodesic flows on closed Riemannian manifolds of negative curvature*, Soviet Math. Dokl. **4** (1963), 1153–1156.
8. M. F. Atiyah and I. M. Singer, *The index of elliptic operators*. III, Ann. of Math. **87** (1968), 546–604.
9. M. F. Atiyah, R. Bott, and V. K. Patodi, *On the heat equation and the index theorem*, Invent. Math. **19** (1973), 279–330.
10. R. Bowen, *Symbolic dynamics for hyperbolic flows*, Amer. J. Math. **95** (1973), 429–460.
11. R. Brooks, *On manifolds of negative curvature with isospectral potentials*, Topology **26** (1987), 63–66.
12. R. Brooks and R. Tse, *Isospectral surfaces of small genus*, Nagoya Math. J. **107** (1987), 13–24.
13. P. Buser, *Isospectral Riemann surfaces*, Ann. Inst. Fourier (Grenoble) **36** (1986), 167–192.
14. J. W. S. Cassels and A. Fröhlich (eds.), *Algebraic number theory*, Academic Press, London and New York, 1967.
15. D. Fried, *The zeta functions of Ruelle and Selberg*. I, Ann. Sci. École Norm. Sup. **19** (1986), 491–517.
16. R. Gangolli, *Zeta functions of Selberg's type for compact space forms of symmetric space of rank one*, Illinois J. Math. **21** (1977), 1–42.
17. F. D. Gantmacher, *Applications of the theory of matrices*, Interscience, New York, 1959.
18. I. M. Gel'fand, *Automorphic functions and the theory of representations*, Proc. Internat. Congress Math., Stockholm, 1962, pp. 74–85.
19. P. B. Gilkey, *Invariance theory, the heat equations, and the Atiyah-Singer index theorem*, Publish or Perish, Houston, Texas, 1985.
20. D. Hejhal, *The Selberg trace formula and the Riemannian zeta function*, Duke Math. J. **43** (1976), 441–482.
21. _____, *The Selberg trace formula for* $PSL(2, \mathbb{R})$, Lecture Notes in Mathematics, vol. 548 and vol. 1001, Springer-Verlag, Berlin and New York.
22. Y. Ihara, *On discrete subgroups of the two by two projective linear group over p-adic fields*, J. Math. Soc. Japan **18** (1966), 219–235.
23. A. Ikeda, *On spherical space forms which are isospectral but not isometric*, J. Math. Soc. Japan **35** (1983), 437–444.
24. M. Kac, *Can one hear the shape of a drum?*, Amer. Math. Monthly **73** (1966), 1–23.
25. A. Katsuda, *Homology of closed geodesics in a nonpositively curved manifold of rank one*, preprint, Nagoya Univ., 1985.
26. A. Katsuda and T. Sunada, *Homology of closed geodesics in certain Riemannian manifolds*, Proc. Amer. Math. Soc. **96** (1986), 657–660.
27. W. Klingenberg, *Riemannian manifolds with geodesic flow of Anosov type*, Ann. of Math. **99** (1974), 1–13.
28. _____, *Lectures on closed geodesics*, Springer-Verlag, New York, 1978.
29. K. Komatsu, *On the adele ring of algebraic number fields*, Kodai Math. Sem. Rep. **28** (1976), 78–84.

30. N. Kurokawa, *On some Euler products.* I, II, Proc. Japan Acad. **60** (1984), 335–338; 365–368.
31. S. Lang, *Algebraic number theory*, Addison-Wesley, Reading, MA, 1970.
32. A. Manning, *Axiom A diffeomorphisms have rational zeta functions*, Bull. London Math. Soc. **3** (1971), 215–220.
33. _____, *Topological entropy for geodesic flows*, Ann. of Math. **110** (1979), 567–573.
34. H. P. McKean, *Selberg's trace formula as applied to a compact Riemann surface*, Comm. Pure Appl. Math. **25** (1972), 225–246.
35. J. J. Millson, *On the first Betti number of a compact negatively curved manifold*, Ann. of Math. **104** (1976), 235–247.
36. J. Milnor, *Eigenvalues of the Laplace operators on certain manifolds*, Proc. Nat. Acad. Sci. USA **51** (1964), 542.
37. S. Minakshisundaram and A. Pleijel, *Some properties of the eigenfunctions of the Laplace operator on Riemannian manifolds*, Canad. J. Math. **1** (1949), 242–256.
38. G. A. Margulis, *Applications of ergodic theory to the investigation of manifolds of negative curvature*, Funct. Anal. Appl. **3** (1969), 335–336.
39. _____, *Discrete groups of motions of manifolds of non-positive curvature*, Proc. Internat. Cong. Math., Vancouver, 1974, vol. 2, pp. 21–34. (Russian).
40. W. Parry and M. Pollicott, *An analogue of the prime number theorem for closed orbits of axiom A flows*, Ann. of Math. **118** (1983), 573–591.
41. _____, *The Chebotarev theorem for Galois coverings of Axiom A flows*, Ergodic Theory Dynamical Systems **6** (1986), 133–148.
42. R. Perlis, *On the equation $\zeta_K(s) = \zeta_{K'}(s)$*, J. Number Theory **9** (1977), 242–260.
43. M. Pollicott, *A complex Ruelle-Perron-Frobenius theorem and two counter-examples*, Ergodic Theory Dynamical Systems **4** (1984), 135–146.
44. _____, *Meromorphic extensions of generalized zeta functions*, Invent. Math. **85** (1986), 147–164.
45. D. Ruelle, *Zeta functions for expanding maps and Anosov flows*, Invent. Math. **34** (1976), 231–242.
46. _____, *Thermodynamic formalism*, Addison-Wesley, Reading, MA, 1978.
47. P. Sarnak, *Prime geodesic theorems*, Ph. D. Dissertation, Stanford Univ., Stanford, CA, 1980.
48. R. T. Seeley, *Complex powers of an elliptic operator*, Proc. Sympos. Pure Math., vol. 10, Amer. Math. Soc., Providence, RI, pp. 288–307.
49. A. Selberg, *Harmonic analysis and discontinuous subgroups in weakly symmetric Riemannian spaces with applications to Dirichlet series*, J. Indian Math. Soc. **20** (1956), 47–87.
50. J.-P. Serre, *Trees*, Springer-Verlag, New York, 1980.
51. Ya Sinai, *The asymptotic behavior of the number of closed orbits on a compact manifold of negative curvature*, Amer. Math. Soc. Transl. Ser. 2 **73** (1967), 227–250.
52. S. Smale, *Differentiable dynamical systems*, Bull. Amer. Math. Soc. **73** (1967), 747–817.
53. T. Sunada, *Spectrum of a compact flat manifold*, Comment. Math. Helv. **53** (1978), 613–621.
54. _____, *Trace formula for Hill's operators*, Duke Math. J. **47** (1980), 529–546.
55. _____, *Trace formula and heat equation asymptotics for non-positively curved manifolds*, Amer. J. Math. **104** (1982), 795–812.
56. _____, *Tchebotarev's density theorem for closed geodesics in a compact locally symmetric space of negative curvature*, preprint.
57. _____, *Trace formulas, Wiener integrals and asymptotics*, Proc. Japan-France Seminar, Spectra of Riemannian Manifolds, Kaigai Publ. Tokyo (1983), 103–113.
58. _____, *Geodesic flows and geodesic random walks*, Geometry of Geodesics and Related Topics, vol. 3, Academic Press, New York, 1984, pp. 47–86.
59. _____, *Riemannian coverings and isospectral manifolds*, Ann. of Math. **121** (1985), 169–186.
60. _____, *Gel'fand's problem on unitary representations associated with discrete subgroups of* $PSL_2(\mathbb{R})$, Bull. Amer. Math. Soc. **12** (1985), 237–238.
61. _____, *Number theoretical analogues in spectral geometry*, Proc. of the 6th Sympos. on Differential Geometry and Differential Equations held at Shanghai, China, 1985, to appear.
62. _____, *L-functions in geometry in some applications*, Proc. of Taniguchi Symposium held at Katata, 1985, to appear.
63. _____, *Trace formulas and the spectra of Laplacians*, Sūgaku **33** (1981), 134–142.
64. T. Sunada and H. Urakawa, *Ray-Singer zeta functions of flat manifolds*, preprint.
65. A. B. Venkov, *Spectral theory of automorphic functions*, Proc. Steklov Inst. Math. **153** (1982).
66. M. F. Vignéras, *Variétés riemanniennes isospectrales et non isométriques*, Ann. of Math. **112** (1980), 21–32.

67. S. Wolpert, *The length spectra as moduli for compact Riemann surfaces*, Ann. of Math. **109** (1979), 323–351.
68. A. Katsuda and T. Sunada, *Homology and closed geodesics in a compact Riemann surface*, Amer. J. Math. **110** (1988), 145–156.
69. C. S. Gordon, D. L. Webb, and S. Wolpert, *One cannot hear the shape of a drum*, Bull. Amer. Math. Soc. **27** (1992), 139–158.
70. C. S. Gordon and E. N. Wilson, *Isospectral deformations of compact solv-manifolds*, J. Differential Geom. **19** (1984), 241–256.
71. D. M. DeTurck and C. S. Gordon, *Isospectral deformations. I*, Comm. Pure Appl. Math. **40** (1987), 367–387.
72. V. Gullemin and D. Kazhdan, *Some inverse spectral results for negatively curved 2-manifolds*, Topology **19** (1980), 301–312.
73. R. Brooks, *On a manifold of negative curvature with isospectral potentials*, Topology **26** (1987), 63–66.
74. A. Katsuda and T. Sunada, *Closed orbits in homology classes*, Inst. Hautes Études Sci. Publ. Math. **71** (1990), 1–32.
75. R. Phillips and P. Sarnak, *Geodesics in homology classes*, Duke Math. J. **55** (1987), 287–297.

DEPARTMENT OF MATHEMATICS, TOHOKU UNIVERSITY, SENDAI 980, JAPAN

Translated by YUICHIRO TAGUCHI

Fundamental Groups and Laplacians

Toshikazu Sunada

Anyone who aspires to be a mathematician has heard of, or encountered in books on the history of mathematics, the celebrated Riemann hypothesis. This hypothesis asserts that for the Riemann zeta function

$$\zeta(s) = \sum_{n=1}^{\infty} n^{-s},$$

which is a meromorphic function on the whole s-plane, all the zeros in the region $0 < \operatorname{Re} s < 1$ lie on the line $\operatorname{Re} s = 1/2$. The Riemann hypothesis still remains open (at this time of writing, April 1987) all the efforts of 20th-century mathematicians notwithstanding. Since 1859—the year Riemann's paper on his zeta function appeared—many programs have been proposed towards a proof of the hypothesis. The most interesting one, due to Hilbert, is based on a rather vague but hopeful conjecture which suggests there is a relation between the zeros of the zeta function and the distribution of the eigenvalues of certain selfadjoint operators. The author and perhaps many of the readers hope that the Riemann hypothesis will be settled in a manner that realizes this vision.

One may ask why it is expected that the eigenvalue problem of selfadjoint operators should arise in connection with the Riemann hypothesis. It is well known that the Riemann zeta function has been conceptually generalized to various zeta functions in algebraic geometry and differential geometry. In fact, some of those zeta functions turn out to be closely related to selfadjoint operators on certain function spaces where the eigenvalues of the operators completely determine the singular points of the zeta functions, and thus, the analogues of the Riemann hypothesis for those zeta functions are reduced to estimates of the eigenvalues. To explain this we consider two examples.

§1. Ihara zeta functions

Let M be a finite regular graph, that is, a finite set such that each vertex $x \in M$ has a neighborhood $V(x) \subset M$ satisfying the following conditions:
  (i) $y \in V(x)$ iff $x \in V(y)$;
  (ii) $x \notin V(x)$;
  (iii) $\# V(x)$ is independent of x (and will be written $q + 1$).

1991 *Mathematics Subject Classification*. Primary 58G25.
This article originally appeared in Japanese in Sūgaku **39** (3) (1987), 193–203.

A pair of vertices $\{x, y\}$ is called an edge if $y \in V(x)$. We obtain a graph in the usual sense by joining x and y by a line segment.

A sequence of vertices $c = (x_0, x_1, \ldots, x_n)$ is called a closed path in M if $\{x_i, x_{i+1}\}$ are edges for all $i = 0, 1, \ldots, n-1$ and $x_0 = x_n$. It is called a closed geodesic if in addition $x_{i-1} \neq x_{i+1}$ for all $i \in \mathbb{Z}/n\mathbb{Z}$. For a closed path $c = (x_0, x_1, \ldots, x_n)$, the k-multiple c^k is the closed path defined by

$$c^k = (\underbrace{x_0, \ldots, x_n}_{1}, \underbrace{x_0, \ldots, x_n}_{2}, x_0, \ldots, x_n, \underbrace{x_0, \ldots, x_n}_{n}).$$

See Figure 1.

A closed path is called prime if it is not a k-multiple of another one ($k \geq 2$). Two closed paths $c_1 = (x_0, x_1, \ldots, x_n)$, $c_2 = (y_0, y_1, \ldots, y_n)$ are said to be equivalent if there exists an integer d (mod n) such that $y_k = x_{k+d}$ ($k \in \mathbb{Z}/n\mathbb{Z}$). An equivalence class of a prime geodesic is called a prime geodesic cycle. If $\mathfrak{p}$ is a cycle represented by a closed path $c = (x_0, x_1, \ldots, x_n)$, then n is called the length of $\mathfrak{p}$ and is denoted by $l(\mathfrak{p})$.

The zeta function of a finite regular graph M is a certain generating function related to $\{l(\mathfrak{p}); \mathfrak{p}$ is a prime geodesic cycle$\}$. As an analogue of the Euler product representation of the Riemann function, we define the zeta function by

$$Z(s) = \prod_{\mathfrak{p}} (1 - u^{l(\mathfrak{p})})^{-1} \qquad (u = q^{-s}).$$

$Z(s)$ converges absolutely in $\operatorname{Re} s > 1$ ($|u| < q^{-1}$) and extends to a meromorphic function of s (rational with respect to u).

We now define a selfadjoint operator corresponding to the zeta function $Z(s)$. Given a regular graph M (not assumed to be finite), we set

$$L^2(M) = \left\{ \psi : M \to \mathbb{C}; \sum_{x \in M} |\psi(x)|^2 < \infty \right\}$$

$L^2(M)$ is a Hilbert space with scalar product

$$\langle \varphi, \psi \rangle = \sum_{x \in M} \varphi(x) \overline{\psi(x)}.$$

An operator $A = A_M : L^2(M) \to L^2(M)$ is defined by

$$(A\varphi)(x) = \sum_{y \in V(x)} \varphi(y).$$

It is easily seen that this is a bounded selfadjoint operator. A is called the *adjacency operator* of the graph M.

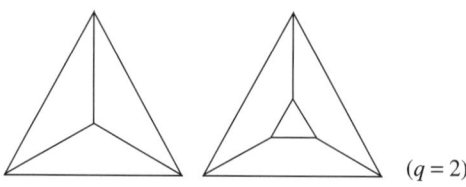

FIGURE 1

If M is a finite graph, then $L^2(M)$ consists of all the functions on M and $q+1$ is a simple eigenvalue of A with constant eigenfunctions. Let $\{\lambda_i\}_{i=0}^{h-1}$ be the eigenvalue of A with

$$q+1 = \lambda_0 > \lambda_1 \geq \lambda_2 \geq \cdots \geq \lambda_{h-1} \geq -(q+1), \qquad h = \#M.$$

PROPOSITION 1. *For a finite regular graph M, we have*

$$Z(s) = (1-u^2)^{-g} \det(I - Au + qu^2)^{-1},$$

where $u = q^{-s}$ and $g = (q-1)h/2$.

REMARK. Ihara [4] defined zeta functions associated with discrete subgroups in $\mathrm{PGL}_2(\mathbb{Q}_p)$ and proved a proposition essentially equivalent to the one above. Serre [8] suggested that Ihara's zeta functions be described in terms of finite graphs, and Sunada gave a concrete shape to this suggestion in his lectures at Nagoya University in 1985. Recently, K. Hashimoto and T. Hori defined zeta functions associated with general algebraic groups of rank 1 over $\mathbb{Q}_p$, which admit an interpretation in terms of two-color graphs.

COROLLARY 1.1. *The singularities (poles) of $Z(s)$ are given as follows*:

$s = \pi n i / \log q, \qquad\qquad n \in \mathbb{Z};$

$s = 1 + 2\pi n i / \log q, \qquad\qquad n \in \mathbb{Z}$ *(if $-(q+1)$ is not an eigenvalue of A)*;

$s = 1 + \pi n i / \log q, \qquad\qquad n \in \mathbb{Z}$ *(if $-(q+1)$ is an eigenvalue of A)*;

$s = -\log u_k / \log q + \pi n i / \log q, \qquad n \in \mathbb{Z};$

$s = 1 + \log u_k / \log q + \pi n i / \log q,$

$$n \in \mathbb{Z} \ (u_k = (\lambda_k + \sqrt{\lambda_k^2 - 4q})/2q, 2q^{1/2} < |\lambda_k| < q+1);$$

$s = (1/2) \pm i t_k + 2\pi n i / \log q,$

$$n \in \mathbb{Z} \ (q^{-1/2 - it_k} = (\lambda_k + \sqrt{\lambda_k^2 - 4q})/2q, |\lambda_k| \leq 2q^{1/2}).$$

Noting that

$$0 < -\log u_k / \log q < 1, \qquad -\log u_k / \log q \neq 1/2, \qquad t_k \in \mathbb{R}$$

we obtain the following corollary.

COROLLARY 1.2. *All the singular points of $Z(s)$ in $0 < \operatorname{Re} s < 1$ lie on the line $\operatorname{Re} s = 1/2$; in other words, $Z(s)$ satisfies the analogue of the Riemann hypothesis if and only if the eigenvalues λ_k except for $\pm(q+1)$ satisfy the estimate $|\lambda_k| \leq 2q^{1/2}$.*

§2. Selberg zeta functions

Let M be a closed Riemann surface with a Riemannian metric of constant curvature $K_M = -1$. Prime closed geodesics in M, as one-dimensional oriented cycles, are called prime geodesic cycles, and $\{\mathfrak{p}\}$ denotes the set of all of them. Let $l(\mathfrak{p})$ denote the length of the curve $\mathfrak{p}$. See Figure 2 on p. 22.

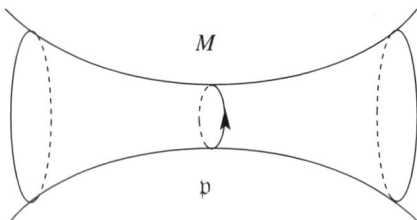

Figure 2

The Selberg zeta function of M is defined by

$$Z(s) = \prod_{k=0}^{\infty} \prod_{\mathfrak{p}} (1 - e^{-(s+k)l(\mathfrak{p})}).$$

It is well known that $Z(s)$ converges absolutely in $\operatorname{Re} s > 1$ and extends to an entire function.

PROPOSITION 2. *$Z(s)$ has zeros at $s = -n$ ($n = 0, 1, 2, \ldots$), and the other zeros are given by*

$$s = 1/2 \pm \sqrt{1/4 - \lambda_k}, \qquad k = 0, 1, 2, \ldots,$$

where

$$0 = \lambda_0 < \lambda_1 \leq \lambda_2 \leq \cdots \uparrow \infty$$

denote the eigenvalues of the Laplacian Δ_M on M. In other words, the singularities (zeros) of the Selberg zeta function are described by the eigenvalues of the selfadjoint operator

$$\Delta_M \colon L^2(M) \to L^2(M).$$

Since $\lambda_n \uparrow \infty$, the above proposition implies that $Z(s)$ "nearly" satisfies the analogue of the Riemann hypothesis in the sense that all the zeros of $Z(s)$ in $0 < \operatorname{Re} s < 1$ lie on the line $\operatorname{Re} s = 1/2$ except for possibly finitely many real zeros. Furthermore,

COROLLARY 2.1. *$Z(s)$ satisfies the analogue of the Riemann hypothesis if and only if*

$$\lambda_1 \geq 1/4.$$

Here is a natural question: what on earth is the nature of the numbers $2q^{1/2}$ and $1/4$ in the estimates in the analogues of the Riemann hypothesis? The answer is given by considering the universal covering spaces of the spaces M.

CASE I. Let

$$\omega \colon X \to M$$

be the universal covering of a finite regular graph (as a one-dimensional simplicial complex) and introduce a graph structure on X such that ω is a morphism of graphs. X is then a regular tree. See Figure 3.

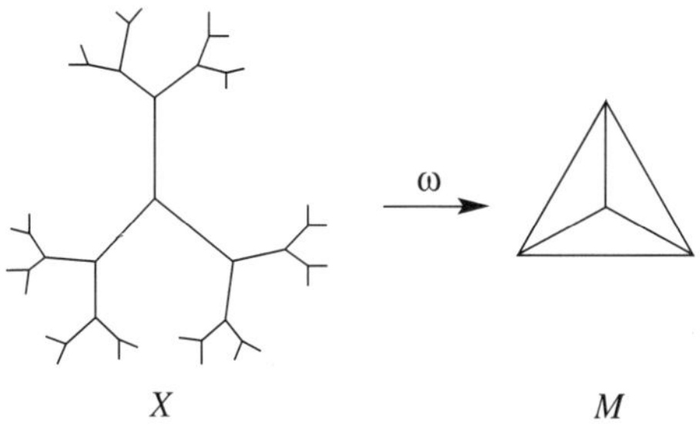

FIGURE 3

Consider the adjacency operator on X, $A\colon L^2(X) \to L^2(X)$. In view of the theory of spherical functions on regular trees, we can obtain a description of the spectrum of A, denoted by $\mathrm{Spect}(A)$.

PROPOSITION 3. $\mathrm{Spect}(A) = [-2q^{1/2}, 2q^{1/2}]$.

In fact it is easy, and left as an exercise to the reader, to show that $\mathrm{Spect}(A) \subset [-2q^{1/2}, 2q^{1/2}]$.

CASE II. The universal covering space of a compact Riemann surface with constant negative curvature ($K_M \equiv 1$) is identified with the upper half-plane with the Poincaré metric ($\mathbb{H}^2, ds^2$):

$$\mathbb{H}^2 = \{(x,y) \in \mathbb{R}^2; y > 0\}, \qquad ds^2 = y^{-2}(dx^2 + dy^2).$$

The Laplacian on $\mathbb{H}^2$ is

$$\Delta_{\mathbb{H}^2} = -y^2\left(\frac{\partial^2}{\partial x^2} + \frac{\partial^2}{\partial y^2}\right)$$

and satisfies, in view of the theory of spherical functions on $\mathbb{H}^2$, the following

PROPOSITION 4. $\mathrm{Spect}(\Delta_{\mathbb{H}^2}) = [1/4, \infty)$.

It is also easy in this case to show $\mathrm{Spect}(\Delta_{\mathbb{H}^2}) \subset [1/4, \infty)$ (exercise).

In both I and II, $2q^{1/2}$ and $1/4$ are the end points of the spectral sets of the operators on the universal covering spaces. These phenomena, especially in Case II, suggest the problem of studying, for a given covering of a compact Riemannian manifold M,

$$\omega\colon X \to M,$$

relationships of the spectrum of the Laplacian Δ_X on X and that of the Laplacian Δ_M on M. Although Case I also suggests studying relationships between the spectrum of the adjacency operator on a finite graph and that of the operator on a covering graph, we consider here only problems concerning Laplacians.

We now establish some terminology. Let (X, ds^2) be a smooth Riemannian manifold. We express the Riemannian metric ds^2 as

$$ds^2 = \sum_{i,j=1}^{n} g_{ij}\, dx_i\, dx_j$$

using local coordinates $(x_1, \ldots, x_n)$. (g_{ij}) is a function valued in symmetric positive matrices. The metric ds^2 determines the length of curves on X and thus defines a distance between two points. (X, ds^2) is called complete if this distance is complete. ds^2 induces, moreover, a canonical measure on X

$$dv = (\det g_{ij})^{1/2} dx_1 \cdots dx_n.$$

$L^2(X)$ denotes the Hilbert space of square integrable functions on X with respect to this canonical measure.

The Laplacian Δ_X acting on functions on M is a second-order partial differential operator defined by

$$\Delta_X = -(\det g_{ij})^{-1/2} \sum_{i,j=1}^n \frac{\partial}{\partial x_i} \left((\det g_{ij})^{1/2} g^{ij} \frac{\partial}{\partial x_j} \right),$$

where (g^{ij}) is the inverse matrix of (g_{ij}). Δ_X is elliptic by the positivity of (g_{ij}), and it is easy to see that

$$\langle \Delta_X f, f \rangle \geq 0 \quad \left(\langle f_1, f_2 \rangle = \int_X f_1 \cdot \overline{f_2} \, dv \right),$$

$$\langle \Delta_X f_1, f_2 \rangle = \langle f_1, \Delta_X f_2 \rangle.$$

Moreover, by the completeness of (X, ds^2), $\Delta_X \colon C_0^\infty(X) \to C_0^\infty(X)$ is extended to a densely defined selfadjoint operator $\Delta_X \colon L^2(X) \to L^2(X)$. A general theory tells us that the spectrum of Δ_X is given by

$$\mathrm{Spect}(\Delta_X) = \left\{ \lambda \in \mathbb{C}; \text{ there is some } \{f_k\}_{k=1}^\infty \subset C_0^\infty(X) \text{ s.t. } \|f_k\| = 1, \right.$$

$$\left. \lim_{k \to \infty} \|\Delta_X f_k - \lambda f_k\| = 0 \right\}.$$

By the nonnegativity of Δ_X, $\mathrm{Spect}(\Delta_X)$ is a closed subset of $[0, \infty)$.

The eigenvalues of Δ_X are elements of $\mathrm{Spect}(\Delta_X)$ by the definition of spectrum. If X is compact, $\mathrm{Spect}(\Delta_X)$ consists only of eigenvalues with finite multiplicities and is discrete in $[0, \infty)$. We write the eigenvalues as

$$0 = \lambda_0(X) < \lambda_1(X) \leq \lambda_2(X) \leq \cdots \uparrow \infty.$$

$\lambda_0(X) = 0$ is the eigenvalue with constant eigenfunctions.

We now return to the problem proposed above. Let $\omega \colon (X, ds_X^2) \to (M, ds_M^2)$ be a *Riemannian covering*, i.e., a covering map that is isometric ($\omega^* ds_M^2 = ds_X^2$). Then we have

$$(\Delta_M f) \circ \omega = \Delta_X (f \circ \omega)$$

for any $f \in C^\infty(M)$. The reader may think at first glance that this relation reduces our problem to an easy problem. In fact, in the case that ω is a finite covering, $f \circ \omega$ belongs to $L^2(X)$ if $f \in L^2(X)$, in other words, if X is compact, and $f \circ \omega$ is an eigenfunction of Δ_X with the same eigenvalue as f if it is an eigenfunction of Δ_M. Thus, we obtain

$$\mathrm{Spect}(\Delta_M) \subset \mathrm{Spect}(\Delta_X).$$

But, we cannot proceed so easily in the case where ω is an infinite covering map.

EXAMPLE 1. The universal covering $X = H^2 \to M$ of a compact Riemann surface M of constant negative curvature (-1) satisfies $0 \in \text{Spect}(\Delta_M)$ but $0 \notin \text{Spect}(\Delta_{H^2})$. Thus,
$$\text{Spect}(\Delta_M) \not\subset \text{Spect}(\Delta_X).$$

EXAMPLE 2. The universal covering of the flat torus $\mathbb{R}^2/\mathbb{Z}^2$ is the Euclidean plane $\mathbb{R}^2$ and satisfies
$$\text{Spect}(\Delta_M) \subset [0, \infty) = \text{Spect}(\Delta_{\mathbb{R}^2}).$$

What causes the difference between these two examples? We realize at once that in Example 1 the covering transformation group (i.e., the fundamental group $\pi_1(M)$ of M) contains the free group F_2, which is far from abelian and in Example 2 the fundamental group $\pi_1(M) = \mathbb{Z}^2$ which is abelian.

The difference between Example 1 and Example 2 is not due to commutativity, but is due to the amenability of the covering transformation groups.

THEOREM A. *Let $\omega: X \to M$ be a normal Riemannian covering, and let G be the covering transformation group of ω. If G is amenable, then we have*
$$\text{Spect}(\Delta_M) \subset \text{Spect}(\Delta_X).$$
In particular, $0 \in \text{Spect}(\Delta_X)$.

Set $\lambda_0(X) = \inf \text{Spect}(\Delta_X)$. Then the condition $0 \in \text{Spect}(\Delta_X)$ is equivalent to $\lambda_0(X) = 0$ ($\lambda_0(X)$ is not necessarily an eigenvalue if X is noncompact). The converse of Theorem A also holds by the following

THEOREM (R. Brooks [2]). *If $\lambda_0(X) = 0$ for a normal Riemannian covering $\omega: X \xrightarrow{G} M$, then G is amenable.*

We review briefly the notion of amenability of a group G (see Greenleaf [3] for details).

DEFINITION. A group G is said to be *amenable* if it has a (left) invariant mean.

An invariant mean on G is a bounded linear functional m on $L_{\mathbb{R}}^\infty(G)$ (the space of $\mathbb{R}$-valued bounded functions on G with the sup norm $\|\cdot\|_\infty$) satisfying the following conditions:

(1) $m(1) = 1$, where 1 on the left-hand side means the function on G is identically equal to 1,

(2) if $f \in L_{\mathbb{R}}^\infty(G)$ is nonnegative, then $m(f) \geq 0$;

(3) if we define $\sigma f \in L_{\mathbb{R}}^\infty(G)$ for $\sigma \in G$, $f \in L_{\mathbb{R}}^\infty(G)$ by
$$(\sigma f)(\mu) = f(\sigma\mu),$$
then $m(\sigma f) = m(f)$.

Next we give some examples.

EXAMPLE A. Finite groups are amenable. In fact, an invariant measure on a finite group G is defined by
$$m(f) = (\sharp G)^{-1} \sum_{\sigma \in G} f(\sigma).$$

EXAMPLE B. Abelian groups are amenable. The existence of invariant measures on them is not trivial. Their construction requires the axiom of choice.

EXAMPLE C. The free groups F_k with k (≥ 2) generators are not amenable.

Generally, if G is amenable and H is a subgroup of G, then H is also amenable. Using this fact, we see that the fundamental groups of Riemann surfaces M with $K_M \equiv -1$ are not amenable. Thus, Theorem A reduces the difference between Example 1 and Example 2 to the amenability of the covering transformation groups.

We have already seen that the analogue of the Riemann hypothesis in the case of the Selberg zeta function associated to a Riemann surface M is equivalent to the estimate of the first positive eigenvalue of Δ_M

$$\lambda_1(M) \geq 1/4 \qquad (= \lambda_0(H^2)).$$

H. P. McKean [5] claimed that this always holds, but his proof contained an error [5½] and B. Randol [6] constructed a counterexample using the Selberg trace formula. The existence of such counterexamples is also deduced from the following general theorem.

THEOREM B (Sunada [9]). *Let $\omega: X \xrightarrow{G} M$ be a normal Riemannian covering, and let G be of infinite order and residually finite, that is, $\bigcap \{H ; H \subset G$ is a subgroup of finite index$\} = (1)$. Then*

$$\inf_{M'} \lambda_1(M') \leq \lambda_0(X),$$

where $M' \to M$ runs through the whole class of finite subcoverings of ω (in particular, M' is also compact).

In particular, $\lambda_0(X) = 0$ if G is amenable, and so we can state Theorem B in that case as follows: for any positive number ε there exists a finite subcovering $M' \to M$ of ω with $\lambda_1(M') < \varepsilon$.

Consider a compact Riemann surface M with $K_M \equiv -1$, and consider the homology universal covering of M, $\omega: X \xrightarrow{G} M$, $G = H_1(M, \mathbb{Z})$. If g (≥ 2) is the genus of M, then G is a free abelian group of rank $2g$, and in particular, G is residually finite. Therefore, there exists a finite subcovering of ω, $M' \to M$, such that $\lambda_1(M') < 1/4$. Since M' is a compact Riemann surface with constant negative curvature, we have a counterexample.

REMARK. The analogue of the Riemann hypothesis which concerns the Selberg zeta functions has deep connections with the distribution of prime geodesics, in the same way as the original Riemann hypothesis has deep connections with the distribution of prime numbers, in particular, with the precise formulations of the prime number theorem.

It is hoped that the above argument has helped the reader to understand that for a given $\omega: X \to M$, the relation between $\mathrm{Spect}(\Delta_X)$ and $\mathrm{Spect}(\Delta_M)$ is not as simple as it may appear. What method is then useful for handling problems of this type? For example, it is trivial that $\mathrm{Spect}(\Delta_M) \subset \mathrm{Spect}(\Delta_X)$ for a finite covering ω, but then what information does $\mathrm{Spect}(\Delta_X) \setminus \mathrm{Spect}(\Delta_M)$ contain? To answer these questions, we introduce the notion of *flat vector bundle*.

Let $\omega: X \to M$ be a normal covering, and let G be the covering transformation group. Consider a unitary representation of G on a Hilbert space V,

$$\rho: G \to U(V) = \{\text{unitary transformation of } V\}.$$

We define the vector bundle E_ρ on M associated with ρ in the following way. Let

G act on $X \times V$ by

$$\sigma(x, v) = (\sigma x, \rho(\sigma)v),$$

and let E_ρ be the quotient space $G\backslash(X \times V)$. If we define the map $p\colon E_\rho \to M$ by $p([x, v]) = \omega(x)$, where $[x, v]$ denotes the orbit of (x, v), then E_ρ has a vector bundle structure with bundle projection p. We call this the flat vector bundle associated with ρ.

We identify the section of E_ρ with the V-valued functions s on X satisfying the conditions:

$$s(\sigma x) = \rho(\sigma) s(x), \qquad \sigma \in G, \quad x \in X.$$

We define the norm of a section s to E_ρ by

$$\left(\int_M \|s(x)\|_V^2 \, dx\right)^{1/2}.$$

Note that $\|s(x)\|_V$ can be considered as a function on M because $\|s(x)\|_V$ satisfies $\|s(\sigma x)\|_V = \|s(x)\|_V$. Let $L^2(E_\rho)$ be the completion of $C^\infty(E_\rho)$ with respect to this norm. Then $L^2(E_\rho)$ has a Hilbert space structure with the natural inner product.

The Laplacian acts on $C^\infty(X, V)$ and, by commutativity with $\sigma \in G$, it maps $C^\infty(E_\rho)$ into itself. If we let Δ_ρ denote the restriction of the Laplacian to $C^\infty(E_\rho)$, then for $s, s_1, s_2 \in C^\infty(E_\rho)$ we have

$$\langle \Delta_\rho s, s \rangle \geq 0, \qquad \langle \Delta_\rho s_1, s_2 \rangle = \langle s_1, \Delta_\rho s_2 \rangle.$$

Moreover, Δ_ρ is extended to a densely defined selfadjoint operator

$$\Delta_\rho \colon L^2(E_\rho) \to L^2(E_\rho).$$

From the properties above of Δ_ρ it follows that $\mathrm{Spect}(\Delta_\rho) \subset [0, \infty)$.

EXAMPLE α. For $V = \mathbb{C}$, $\rho = 1$ (the trivial representation), we have

$$L^2(E_\rho) \cong L^2(M), \quad \Delta_\rho \cong \Delta_M. \quad \text{(unitarily equivalent)}$$

EXAMPLE β. Let $V = L^2(G)$, and let $\rho = \rho_r$ be the *right regular representation*; that is,

$$(\rho_r(\sigma)\varphi)(\mu) = \varphi(\mu\sigma)$$

for $\varphi \in L^2(G)$, $\sigma, \mu \in G$. Then we have

$$L^2(E_\rho) \cong L^2(X), \quad \Delta_\rho \cong \Delta_X. \quad \text{(unitarily equivalent)}$$

In fact, if we define a function $s\colon X \to L^2(G)$ by

$$s(x)(\sigma) = f(\sigma x)$$

for $f \in C_0^\infty(X)$, then the correspondence $f \mapsto s$ extends to the unitary equivalence we need.

Let $\omega\colon X \xrightarrow{G} M$ be a finite regular covering. Corresponding to the irreducible decomposition of the normal representation of the finite group G

$$\rho_r = \sum_{\rho \in \widehat{G}} \oplus (\dim \rho)\rho$$

(where $\widehat{G}$ is the set of the equivalence classes of irreducible representations) we have a decomposition of the Laplacian Δ_{ρ_r}

$$\Delta_{\rho_r} = \sum_{\rho \in \widehat{G}} \oplus (\dim \rho)\Delta_\rho.$$

Thus, we obtain the following proposition. (This is an answer to the problem posed above.)

PROPOSITION 5.
$$\mathrm{Spect}(\Delta_X) = \bigcup_{\rho \in \widehat{G}} \mathrm{Spect}(\Delta_\rho).$$

We now return to the general case. First, consider the minimum spectrum of Δ_ρ. For simplicity we set

$$\lambda_0(\rho) = \inf \mathrm{Spect}(\Delta_\rho) \qquad (\geq 0)$$

and consider how to estimate $\lambda_0(\rho)$ from above and below by purely representation-theoretic quantities.

DEFINITION. Let $A \subset G$ be a finite system of generators and set

$$\delta(\rho, 1) = \delta_A(\rho, 1) = \inf_{\substack{v \in V \\ \|v\|=1}} \sup_{\sigma \in A} \|\rho(\sigma)v - v\|_V.$$

$\delta(\rho, 1)$ is something like a "distance" between ρ and the trivial representation 1.

THEOREM C. *There exist positive constants C_1, C_2 independent of the representation ρ such that*

$$C_1 \delta(\rho, 1)^2 \leq \lambda_0(\rho) \leq C_2 \delta(\rho, 1)^2$$

holds for every representation ρ.

COROLLARY. (i) $\lambda_0(\rho) = 0$ iff $\delta(\rho, 1) = 0$.
(ii) *The following two conditions are equivalent*:

"*there exists a positive constant C such that $\delta(\rho, 1) \geq C$ holds for every irreducible unitary representation*";

"*there exists a positive contant C' such that $\lambda_0(\rho) \geq C'$ holds for every irreducible unitary representation*".

The former condition in part (ii) of the corollary says that G satisfies the *Kazhdan property* (T). Part (i) of the corollary implies the theorem of Brooks as a special case, since we know the following fact ([15]).

LEMMA. $\delta(\rho_r, 1) = 0$ iff G is amenable.

By part (ii) of the corollary and Proposition 5, we obtain

THEOREM D. *If the covering transformation group G of a regular covering $\omega \colon X \xrightarrow{G} M$ satisfies the Kazhdan property* (T), *then there exists a positive constant C such that*

$$\lambda_1(M') \geq C$$

for any finite subcovering $M' \to M$ of ω.

REMARK. The fundamental groups of compact locally symmetric spaces of rank ≥ 2 satisfy the Kazhdan property (T).

We reconsider Theorem A from the viewpoint of flat bundles. For this purpose we introduce the notion of *weak containment* of representations ([15]). First, we note the trivial fact that $\mathrm{Spec}(\Delta_{\rho_1}) \subset \mathrm{Spec}(\Delta_\rho)$ if ρ_1 a subrepresentation of ρ. We now see that this still holds if we replace "subrepresentation" by a weaker notion.

Let ρ_1, ρ be unitary representations of G on Hilbert spaces V, V_1, respectively. We say that ρ_1 is weakly contained in ρ (written as $\rho_1 \prec \rho$) if for any positive constant ε and any orthogonal basis $\{v_1, \ldots, v_n\}$ of V_1 there exists an orthogonal basis $\{u_1, \ldots, u_n\}$ of V such that

$$|\langle v_i, \rho_1(\sigma) v_j \rangle_{V_1} - \langle u_i, \rho(\sigma) u_j \rangle_V| < \varepsilon$$

holds for every $\sigma \in A$, $i, j = 1, \ldots, n$. It is easily checked that $1 \prec \rho$ iff $\delta(\rho, 1) = 0$.

THEOREM E. *If $\rho_1 \prec \rho$, then $\mathrm{Spect}(\Delta_{\rho_1}) \subset \mathrm{Spect}(\Delta_\rho)$.*

Theorem A follows from this theorem. In fact, if G is amenable, then any representation ρ of G is weakly contained in an infinite sum of (copies of) a regular representation ρ_r ($\rho \prec \infty \rho_r$). Therefore, $\mathrm{Spect}(\Delta_\rho) \subset \mathrm{Spect}(\Delta_{\rho_r}) = \mathrm{Spect}(\Delta_X)$. In particular, if ρ is the trivial representation 1, then we have $\mathrm{Spect}(\Delta_M) = \mathrm{Spect}(\Delta_1) \subset \mathrm{Spect}(\Delta_X)$ (see Examples α and β).

For a finite-dimensional representation ρ, E_ρ is a vector bundle of finite rank and Δ_ρ is an elliptic differential operator in the usual sense; thus, $\mathrm{Spect}(\Delta_\rho)$ consists only of eigenvalues with finite multiplicities. In particular, $\lambda_0(\rho)$ is an eigenvalue and

$$\lambda_0(\rho) = 0 \quad \text{iff} \quad \rho \text{ contains 1 as a subrepresentation.}$$

Consider the special case of the one-dimensional representation (written as χ) of $G = \pi_1(M)$. The character group of the fundamental group

$$\hat{\pi}_1 = \{\chi \colon \pi_1(M) \to U(1)\}$$

is isomorphic to the character group $\widehat{H}_1$ of the homology group $H_1(M, \mathbb{Z})$ ($= \pi_1/[\pi_1, \pi_1]$) and the identity component of $\widehat{H}_1$ is identified with the torus

$$A(M) = H^1(M, \mathbb{R})/H^1(M, \mathbb{Z}).$$

If we consider Theorem C in the case of one-dimensional representations, then we obtain information about the behavior of the eigenvalue $\lambda_0(\chi)$ near $\chi = 1$, but we can get more precise information by the following argument.

By the Hodge-Kodaira theory, we can identify $H^1(M, \mathbb{R})$ with the space of harmonic 1-forms on M

$$\{\omega; d\omega = \delta\omega = 0\}.$$

We define the character $\chi_\omega \in \hat{\pi}_1$ for $\omega \in H^1(M, \mathbb{R})$ by

$$\chi_\omega(\sigma) = \exp\left(2\pi\sqrt{-1}\int_{C(\sigma)} \omega\right),$$

where $C(\sigma)$ denotes a closed curve whose homotopy class is $\sigma \in \pi_1(M)$. Note that the correspondence $\omega \mapsto \chi_\omega$ gives an isomorphism from $A(M)$ to the identity component of $\hat{H}_1$.

Set $\lambda_0(\omega) = \lambda_0(\chi_\omega)$. Then $\lambda_0(\omega)$ is a continuous function on $H^1(M, \mathbb{R})$ with $\lambda_0(0) = 0$ and $\lambda_0(\omega) \geq 0$.

On the other hand, $H^1(M, \mathbb{R})$ is considered as a Euclidean space with norm

$$\|\omega\| = \left(\int_M |\omega|^2 \, dv\right)^{1/2}.$$

Because $\lambda_0(0)$ is a simple eigenvalue, $\lambda_0(\omega)$ is also simple for any ω with sufficiently small $\|\omega\|$, and therefore, we can apply perturbation theory to obtain the following theorem.

THEOREM F. *$\lambda_0(\omega)$ is of class C^∞ and satisfies*

$$\lambda_0(\omega) = \frac{4\pi^2}{\text{vol}(M)}\|\omega\|^2 + O(\|\omega\|^4).$$

Moreover, we have

$$\lambda_0(\omega) \leq \frac{4\pi^2}{\text{vol}(M)}\|\omega\|^2$$

for every $\omega \in H^1(M, \mathbb{R})$.

We apply the above theorem to the problem of the distribution of prime geodesic cycles on a Riemann surface. Set

$$\pi(x, \alpha) = \#\{\mathfrak{p} \colon \text{a prime geodesic cycle}; l(\mathfrak{p}) < x, [\mathfrak{p}] = \alpha\}$$

for a homology class $\alpha \in H_1(M, \mathbb{R})$ of a compact Riemann surface M with constant negative curvature. The following theorem is a geometric analogue of the prime number theorem on arithmetical progressions.

THEOREM G (Katsuda-Sunada [14]). *Let g be the genus of M. Then we have*

$$\pi(x, \alpha) \sim (g-1)^g e^x / x^{g+1} \qquad (x \uparrow \infty),$$

where $f(x) \sim h(x)$ $(x \uparrow \infty)$ means $\lim_{x \to \infty} f(x)/h(x) = 1$.

We give a sketch of a proof. We consider a slight modification of the zeta function defined in II,

$$Z(s,\chi) = \prod_{k=0}^{\infty}\prod_{\mathfrak{p}}(1 - \chi([\mathfrak{p}])e^{-(s+k)l(\mathfrak{p})}).$$

In the same way as $Z(s) = Z(s,1)$, $Z(s,\chi)$ is extended over the whole plane and its zeros in $0 \leq \operatorname{Re} s \leq 1$ are given by

$$\tfrac{1}{2} \pm \sqrt{\tfrac{1}{4} - \lambda_k(\chi)}, \qquad k = 0, 1, 2, \ldots,$$

where $0 \leq \lambda_0 \leq \lambda_1 \leq \cdots$ denote the eigenvalues of Δ_χ. Set

$$s(\chi) = \tfrac{1}{2} + \sqrt{\tfrac{1}{4} - \lambda_0(\chi)} \quad \text{and} \quad L(s,\chi) = Z(s+1,\chi)/Z(s,\chi);$$

then there exists a neighborhood of 1, $U \subset \hat{\pi}_1$, and a positive number δ ($< 1/2$) such that

$$(*) = \frac{L'(s,\chi)}{L(s,\chi)} + \frac{1}{s - s(\chi)}, \qquad \chi \in U$$

is regular in the domain $\operatorname{Re} s > 1 - \delta$. We define $F_\alpha(s)$ by

$$F_\alpha(s) = -\int_{A(M)} \chi(-\alpha)\left(-\frac{d}{ds}\right)^g \frac{L'(s,\chi)}{L(s,\chi)}\, d\chi,$$

where $d\chi$ is a normalized Haar measure on $A(M)$. Then by the properties of $(*)$ and Theorem F there exists a locally integrable function $h(t)$ that satisfies, for any sufficiently small $\varepsilon > 0$,

$$|F_\alpha(s) - (g-1)^g/(s-1)| \leq h(t),$$

where $s = 1 + \varepsilon + \sqrt{-1}t$. On the other hand, by applying the orthogonal relations of characters we have

$$F_\alpha(s) = \sum_{\substack{k=1 \\ k[\mathfrak{p}]=\alpha}}^{\infty} \sum_{\mathfrak{p}} k^g l(\mathfrak{p})^{g+1} e^{-skl(\mathfrak{p})};$$

thus, by a Tauberian theorem we see that

$$\varphi(x) = \sum_{\substack{k=1 \\ kl(\mathfrak{p})<x \\ k[\mathfrak{p}]=\alpha}}^{\infty} \sum_{\mathfrak{p}} k^g l(\mathfrak{p})^{g+1}$$

satisfies $\varphi(x) \sim (g-1)^g x$. By using a slight modification of the proof of the prime number theorem, we obtain Theorem G.

REMARK. Theorem G was also proved independently by Phillips and Sarnak.

The reader might be disappointed with the development of our story. We started with a big vision which embraces a proof of the Riemann hypothesis and ended up in a small world of the author's own mathematics. I hope, however, that at the very least my motivation has been clearly explained.

References

1. M. F. Atiyah, *Elliptic operators, discrete groups and von Neumann algebras*, Astérisque **32–33** (1976), 43–72.
2. R. Brooks, *The fundamental groups and the spectrum of the Laplacian*, Comment. Math. Helv. **56** (1981), 581–598.
3. F. P. Greenleaf, *Invariant means on topological groups and their applications*, van Nostrand, Princeton, N.J., 1969.
4. Y. Ihara, *On discrete subgroups of the two by two projective linear group over a p-adic field*, J. Math. Soc. Japan **18** (1966), 219–235.
5. H. P. McKean, *Selberg's trace formula as applied to a compact surface*, Comm. Pure Appl. Math. **25** (1972), 225–246.
$5\frac{1}{2}$. _____, *Correction to* [5], Comm. Pure Appl. Math. **27** (1974), 134.
6. B. Randol, *Small eigenvalues of the Laplace operators on compact Riemann surfaces*, Bull. Amer. Math. Soc. **80** (1974), 996–1000.
7. A. Selberg, *Harmonic analysis and discontinuous subgroups in weakly symmetric Riemannian spaces with applications to Dirichlet series*, J. Indian Math. Soc. **20** (1956), 47–87.
8. J.-P. Serre, *Trees*, Springer-Verlag, New York, 1980.
9. T. Sunada, *Riemannian coverings and isospectral manifolds*, Ann. of Math. **121** (1985), 169–189.
10. _____, *L-functions in geometry and some applications*, Proceeding of Taniguchi Symposium (1985), Lecture Notes in Mathematics, vol. 1201, Springer-Verlag, Berlin and New York, 266–284, 1986.
11. _____, *Unitary representations of fundamental groups and the spectrums of twisted Laplacians*, Topology **28** (1989), 125–132.
12. _____, *On number-theoretic methods in geometry: the geometric analogues of zero- and L-functions and their applications*, Sūgaku **38** (1986), 289–301 (Japanese).
13. _____, *Fundamental Groups and Laplacians*, Kinokuniya, Tokyo, 1988 (Japanese).
14. A. Katsuda and T. Sunada, *Homology and closed geodesics in a compact Riemann surface*, Amer. J. Math. **110** (1988), 145–155.
15. R. J. Zimmer, *Ergodic theory and semisimple groups*, Birkhäuser, Boston, 1984.

DEPARTMENT OF MATHEMATICS, TOHOKU UNIVERSITY, SENDAI 980, JAPAN

Translated by TAKASHI OTUFUJI

Moonshine: A Mysterious Relationship between Simple Groups and Automorphic Functions

Masao Koike

At Santa Cruz in 1979, precisely at the time when the classification of simple groups was about to be completed, Conway and Norton announced that there is a previously unknown field in the study of sporadic simple groups.

Conway and Norton conjectured that to each conjugacy class of the sporadic simple group M of largest order, called the Monster, there corresponds an automorphic function satisfying various amazing relations. This unexpected connection between the two fields came as a surprise. Their paper "Monstrous Moonshine" is full of formulas and relations without any explanation of how they obtained them. It appears that this totality, which certainly contains something in it but has not yet revealed itself in its clear form, was poetically dubbed "moonshine".[1]

The conjecture was based upon the enormous character table[2] constructed by using a computer and by assuming a certain conjecture[3] on the Monster, the existence proof of which was at that time not yet in existence. Today we possess an existence proof of M due to Griess, but a complete proof of the moonshine conjecture has not yet been obtained.

In the present note, we describe this extremely interesting conjecture and its relationship to automorphic forms as revealed—though hardly under the sunlight yet—from the standpoint of automorphic function theory.

§1. Automorphic functions and automorphic forms

$\mathrm{GL}_2^+(\mathbb{R}) = \{\sigma = \begin{pmatrix} a & b \\ c & d \end{pmatrix} \mid \det \sigma > 0\}$ acts on the upper half-plane $\mathfrak{H} = \{z \in \mathbb{C} \mid \mathrm{Im}\, z > 0\}$ as a group of linear fractional transformations

$$z \mapsto \sigma z = \frac{az+b}{cz+d}.$$

$\mathrm{GL}_2^+(\mathbb{R})$ acts also on $\mathbb{R} \cup \{\infty\}$ in the same way.

1991 *Mathematics Subject Classification.* Primary 20D08.
This article originally appeared in Japanese in Sūgaku **40** (3) (1988), 237–246.

[1] No definition of the word moonshine has been given in [**CN**]. The reason why it was so named is given in [**C1**] and the actual place where the discovery was made is also noted. [**C1**] is interesting reading.
[2] Refer to [Atlas].
[3] The conjecture is that M has an irreducible representation of degree 196883.

For each natural number N, consider the congruence subgroup

$$\Gamma_0(N) = \left\{ \begin{pmatrix} a & b \\ c & d \end{pmatrix} \in \mathrm{SL}_2(\mathbb{Z}) \,|\, c \equiv 0 \pmod{N} \right\}.$$

All the groups that we consider hereafter are those groups Γ containing $\Gamma_0(N)$ as a subgroup of finite index for some N.

A nonidentity element $\gamma \in \Gamma$ is termed parabolic if it has only one fixed point in $\mathbb{R} \cup \{\infty\}$. Such a fixed point is called a cusp of Γ. If we set

$$\mathfrak{H}^* = \mathfrak{H} \cup \{\text{all cusps of } \Gamma\},$$

then the quotient space $\Gamma \backslash \mathfrak{H}^*$ has the structure of a closed Riemann surface. We then write X_Γ for this closed Riemann surface and K_Γ for the field of all meromorphic functions on X_Γ. The genus of X_Γ is called the genus of Γ. An element of K_Γ, when pulled back to a function on $\mathfrak{H}$, is called an automorphic function with respect to Γ.

Since the automorphic functions that arise in moonshine are mysteriously only those with respect to some Γ of genus 0, we will assume that

(1) the genus of Γ is 0.

Since $\begin{pmatrix} 1 & 1 \\ 0 & 1 \end{pmatrix} \in \Gamma_0(N)$, ∞ is a cusp of Γ. We assume the stabilizer $\Gamma_\infty = \{\gamma \in \Gamma \,|\, \gamma\infty = \infty\}$ of ∞ in Γ satisfies

(2) $\pm \Gamma_\infty = \pm \{\begin{pmatrix} 1 & n \\ 0 & 1 \end{pmatrix} \,|\, n \in \mathbb{Z}\}$.

Under these two assumptions, an element f of K_Γ that fulfills the following conditions is uniquely determined up to a constant term.

(i) The Fourier expansion of f in a local parameter $q = e^{2\pi i z}$ around ∞ can be written

$$f(z) = q^{-1} + \sum_{n \geq 0} a_n q^n;$$

(ii) f is regular on $\mathfrak{H}$ and also at every cusp of Γ that is not Γ-equivalent to ∞.

The function f even becomes a generator of K_Γ. This f is called the type belonging to Γ.

EXAMPLE. $\Gamma = \mathrm{SL}_2(\mathbb{Z})$ has genus 0, and the type belonging to Γ is the elliptic modular function $J(z)$, whose Fourier expansion at ∞ can be written

$$J(z) = q^{-1} + 744 + 196884q + 21493760q^2 + \cdots.$$

Let us try to write concretely the Γ of genus 0 arising in moonshine. An element $W_{Q,N}$ of the normalizer of $\Gamma_0(N)$ is termed an Atkin-Lehner involution provided that

$$W_{Q,N} = \begin{pmatrix} aQ, & b \\ cN, & dQ \end{pmatrix}; \quad a, b, c, d \in \mathbb{Z}, \ \det W_{Q,N} = Q,$$

where Q is a Hall divisor of N, i.e., a positive divisor of N satisfying $(Q, \frac{N}{Q}) = 1$ we write $\Gamma_0(N)+$ for the group obtained from $\Gamma_0(N)$ by adjoining $W_{Q,N}$ for all Hall divisors Q of N.

Then
$$\Gamma_0(N) + /\Gamma_0(N) \cong \underbrace{(2,\ldots,2)}_{r},$$

where r is the number of distinct prime factors of N. For each Γ such that $\Gamma_0(N)+ \supset \Gamma \supset \Gamma_0(N)$, we write
$$S = \{Q \mid Q \text{ is a Hall divisor of } N \text{ such that } W_{Q,N} \in \Gamma\}$$
to denote $\Gamma = N + S$. In particular, $\Gamma_0(N)+$ and $\Gamma_0(N)$ are denoted $N+$ and $N-$, respectively.

All the groups of the form $N + S$ having genus 0 are known.

If these were all the groups to be dealt with, then the situation would be simple and straightforward. However, we have to consider, for N and its divisor h, the conjugate
$$\begin{pmatrix} h & 0 \\ 0 & 1 \end{pmatrix}^{-1} \cdot \left(\frac{N}{h} + S\right) \cdot \begin{pmatrix} h & 0 \\ 0 & 1 \end{pmatrix}$$
of the group $N/h + S$. We write $\Gamma(N/h + S; h)$ for a certain subgroup of index h in the above conjugate group. We refer to [**Koi6**] for the full details of the definition.

Next, we give a brief review of automorphic forms.

Let χ be a Dirichlet character mod N, let k be an integer such that $k \geq 1$ and $\chi(-1) = (-1)^k$. A function $f(z)$ on $\mathfrak{H}$ is then called an automorphic form on $\Gamma_0(N)$ of weight k and character χ provided that the following conditions are satisfied:

(i) $f(z)$ is regular on $\mathfrak{H}$, and for every $\gamma = \begin{pmatrix} a & b \\ c & d \end{pmatrix} \in \Gamma_0(N)$ the transformation formula
$$f(\gamma z) = \chi(d)(cz + d)^k f(z)$$
holds;

(ii) $f(z)$ is regular at every cusp of Γ.

We write $M_k(N, \chi)$ for the space of the automorphic forms on $\Gamma_0(N)$ of weight k and character χ. An automorphic form $f(z)$ is called a cusp form if it is 0 at every cusp of Γ. We write $S_k(N, \chi)$ for the subspace formed by the cusp forms. Those automorphic forms that fill the gap between $M_k(N, \chi)$ and $S_k(N, \chi)$ are Eisenstein series. Their Fourier coefficients, unlike those of cusp forms, can be written explicitly. We write $\mathscr{E}_k(N, \chi)$ for the subspace formed by the Eisenstein series. We have
$$M_k(N, \chi) = \mathscr{E}_k(N, \chi) \oplus S_k(N, \chi).$$

For a formal Laurent series $\sum a_n q^n$ in q, we define the linear operators U_p and V_p as follows:
$$\sum a_n q^n | U_p = p \sum a_{np} q^n, \qquad \sum a_n q^n | V_p = \sum a_n q^{np}.$$

For a prime p, we define the Hecke operator T_p by $T_p = U_p$ when $p | N$, and
$$f(z)|T_p = p^{-1} f(z)|U_p + p^{k-1} \chi(p) f(z)|V_p$$
when $p \nmid N$. T_p operates on each of $\mathscr{E}_k(N, \chi)$ and $S_k(N, \chi)$. In the study of automorphic forms, it was a cusp form, that is a common eigenfunction for all Hecke operators, that brought various interesting results to number theory in connection with algebraic geometry in particular. As we shall see in §§3 and 4, such a cusp form appears in connection with simple groups as well. Is there any significance in all this?

§2. Main conjecture

The main conjecture of Conway and Norton claims that to each conjugacy class $\langle m \rangle$ of the Monster M there corresponds a formal power series (called a Thompson series)

$$T_m(z) = q^{-1} + 0 + \sum_{n \geq 1} H_n(m) q^n$$

satisfying the following conditions.

(M0) If e is an identity element of M, $T_e(z) = J(z) - 744$.

(M1) If the order of m is n, then there exists some integer h satisfying $h | (24, n)$ and there exists some $\Gamma = \Gamma(n|h + S; h)$ containing $\Gamma_0(nh)$ such that

(i) the genus of Γ is 0,

(i) $T_m(z)$ is the type belonging to Γ.

(M2) For each $n \geq 1$, $H_n: m \mapsto H_n(m)$ is a character of M.

Although it can be easily understood that "automorphic functions of genus 0" and "the Fourier coefficients forming group characters" are the main features of the conjecture, some remarks on the background seem appropriate.

From (M0) and (M2) it follows immediately that the Fourier coefficient of q^n ($n \geq 1$) in $J(z)$ can be expressed as a sum of degrees of irreducible representations of M. But it was McKay who first noticed that the coefficient 196884 of q can be written as the sum $1 + 196883$ of the degree of the trivial representation of M and that of the nontrivial irreducible representation of least degree of M.

Then Thompson ascertained the occurrence of similar phenomena for the Fourier coefficients of $q^2, q^3, \ldots$ and further suggested that inquiries should be made as to what functions would result if formal power series were constructed by letting the characters of the representations thus determined assume the values on various nonidentity elements of M, and the proposal led Conway and Norton to the discovery of (M1).

It was Ogg who noted that the following two conditions for a prime p are equivalent:

(i) the genus of the group $p+$ is 0;

(ii) p divides the order of M.

Indeed, the Thompson series that corresponds to the conjugacy class of M whose Atlas name is pA is the type belonging to $p+$.

In [CN] several pieces of information about $T_m(z)$ are given: the Fourier coefficients up to q^{10}, the η-product expressions, and so on. In view of these, it can be proved that for each conjugacy class $\langle m \rangle$, $T_m(z)$ satisfying (M1) is actually concretely determined. From this viewpoint, therefore, the difficult part of the conjecture is the assertion (M2). See [S] for an attempt at a proof.[4]

In order to understand moonshine, we investigate whether other finite groups have moonshine properties with the conditions (M1) and (M2) slightly relaxed.

A finite group G is said to have moonshine properties whenever to each conjugacy class $\langle g \rangle$ of G there corresponds a formal power series

$$j_g(z) = q^{-1} + \sum_{n \geq 0} a_n(g) q^n$$

---

[4] That H_n is indeed a generalized character of M is a theorem of Atkin-Fong-Smith-Thompson, which was proved by reducing it to congruences among automorphic functions by using a theorem of Brauer-Thompson. It has not been proved that it is a character.

satisfying the following conditions (M1)′ and (M2)′:

(M1)′ $j_g(z)$ is the type belonging to some group Γ of genus 0 containing $\Gamma_0(N)$;

(M2)′ for each $n \geq 1$, $a_n \colon g \mapsto a_n(g)$ is a generalized character of G.

In §§4 and 5, we describe two methods of constructing moonshine for groups smaller than the Monster.

§3. Frame shapes and η-products

As in Table 3 of [CN], there are many cases in which a Thompson series can be written as a product of Dedekind η-functions:

$$\eta(z) = q^{1/24} \prod_{n=1}^{\infty} (1-q^n), \qquad q = e^{2\pi i z}.$$

Let us introduce a Frame shape to explain this.

We call

$$\pi = \prod_{1 \leq t \in \mathbb{Z}} t^{r_t} = 1^{r_1} \cdot 2^{r_2} \cdots, \qquad r_t \in \mathbb{Z}$$

a generalized permutation if $r_t = 0$ except for a finite number of t. This is an extension of the correspondence of a permutation of degree n to the formal product of the lengths of cycles in the canonical cycle product expression.

EXAMPLE. $(12)(34)(567)(8) \mapsto 1 \cdot 2^2 \cdot 3$.

We define the degree and weight of π by

$$\deg \pi = \sum t \cdot r_t, \qquad \text{wt } \pi = \frac{1}{2} \sum r_t.$$

For each generalized permutation $\pi = \prod t^{r_t}$, we define its η-product $\eta_\pi(z)$ by

$$\eta_\pi(z) = \prod \eta(tz)^{r_t}.$$

It is known that $\eta(\sigma z)$ has a transformation formula of "weight $\frac{1}{2}$" which is valid for each element σ of $\mathrm{SL}_2(\mathbb{Z})$.

Consequently, when π satisfies some appropriate conditions—combinatorial or group theoretic—it can be proved that $\eta_\pi(z)$ becomes an automorphic form of weight wt π as described in §1. For example, we mention the following theorem ([DKM], [Koi1]) which was initially discovered by utilizing a computer.[5]

THEOREM 1. *Let $\pi = \prod t^{r_t}$ be a generalized permutation such that $r_t \geq 0$ for all t and wt $\pi \in \mathbb{Z}$, and let t_l and t_s be the largest and smallest t such that $r_t \neq 0$, respectively. Then the following conditions* (A) *and* (B) *are equivalent:*

---

[5] The η-product of a generalized permutation π of degree 24 is said to be multiplicative provided that when $\eta_\pi(z) = \sum_{n \geq 1} a_n q^n$, $(n, m) = 1$ implies $a_{nm} = a_n a_m$. In [DKM], all of the π's have been found whose η-products are multiplicative when $\pi = \prod t^{r_t}$ satisfies the condition that $r_i \geq 0$ for all i. In [Kon2], all of the π's have been found whose η-products are multiplicative when $\pi = \prod t^{r_t}$ satisfies the condition that $\prod(x^t - 1)^{r_t}$ is a polynomial. Though all Frame shapes of $\cdot O$ of type C and all those of type E are of course included, there are many others, and their relation to finite groups is as yet not clear. It is unknown yet whether the multiplicative π's are finite in number if the polynomial condition is dropped.

(A) $\eta_\pi(z)$ is a cusp form, indeed a new form,[6] and is a common eigenfunction for all Hecke operators;

(A) π satisfies the following three conditions:
   (i) if $r_t \neq 0$, then t is a divisor of t_1;
   (ii) setting $N = t_1 t_s$, $\prod t^{r_t} = \prod (\frac{N}{t})^{r_t}$;
   (iii) $\deg \pi = 24$.

Generalized permutations π satisfying condition (B) are given explicitly in Table 1.

When we compare Table 1 with the tables of characters of finite groups in *The Encyclopedic Dictionary of Mathematics* edited by the Mathematical Society of Japan (MIT Press, 1977 and 1987), we observe that our table contains, mysteriously, all the permutation expressions of the elements of the Mathieu group M_{24},[7] and Table 1 is also related to the Conway group $\cdot O$,[8] as we describe in the next section.

Whenever there is a d-dimensional rational representation

$$\rho: G \to \mathrm{GL}_d(\mathbb{Q})$$

of a finite group G, the characteristic polynomial $\det(XI_d - \rho(g))$ of $\rho(g)$ can be written uniquely in the form

$$\prod_{1 \leq t \in \mathbb{Z}} (X^t - 1)^{r_t}, \qquad r_t \in \mathbb{Z}.$$

Hence, a generalized permutation $\pi_g = \prod_{1 \leq t \in \mathbb{Z}} t^{r_t}$ is uniquely determined. We call this permutation the Frame shape of g with respect to ρ. Since ρ is in many cases determined in advance, we simply write $\eta_g(z)$ for the η-product of π_g. The degree of π_g is d.

If we take the symmetric group S_n of degree n and consider its permutation representation of degree n, then all the partitions of n appear as Frame shapes and,

TABLE 1

wt π	π
1	$23 \cdot 1$, $22 \cdot 2$, $21 \cdot 3$, $20 \cdot 4$, $18 \cdot 6$, $16 \cdot 8$, 12^2
2	$15 \cdot 5 \cdot 3 \cdot 1$, $14 \cdot 7 \cdot 2 \cdot 1$, $12 \cdot 6 \cdot 4 \cdot 2$, $11^2 \cdot 1^2$, $10^2 \cdot 2^2$, $9^2 \cdot 3^2$, $8^2 \cdot 4^2$, 6^4
3	$8^2 \cdot 4 \cdot 2 \cdot 1^2$, $7^3 \cdot 1^3$, $6^3 \cdot 2^3$, 4^6
4	$6^2 \cdot 3^2 \cdot 2^2 \cdot 1^2$, $5^4 \cdot 1^4$, $4^4 \cdot 2^4$, 3^8
5	$4^4 \cdot 2^2 \cdot 1^4$
6	$3^6 \cdot 1^6$, 2^{12}
8	$2^8 \cdot 1^8$
12	1^{24}

---

[6] Let χ be defined modulo M, where M is a positive proper divisor of N. For a positive divisor d of N/M, $S_k(M, \chi)^d = \{f(dz) \mid f \in S_k(M, \chi)\}$ is a subspace of $S_k(N, \chi)$, and the space spanned by all of these is called the space of the old forms. Its orthogonal complementary space relative to the Petersson inner product is called the space of the new forms. This is also closed under the action of Hecke operators.

[7] Well-known cusp forms of weight 2 such as $\eta(11z)^2 \eta(z)^2$ are included among them, but why have researchers in automorphic forms not noticed this fact first?

[8] However, among those π satisfying (B), $9^2 \cdot 3^2$, $18 \cdot 6$, and $16 \cdot 8$ do not arise in $\cdot O$ either.

hence, a symmetry of π as seen in condition (B) of Theorem 1 does not hold for every element.

Consequently, the corresponding $\eta_g(z)$ cannot be declared to be interesting. The same situation also takes place in the case of the automorphism group of the E_8-lattice. The group is too large. It appears, therefore, that moonshine cannot be constructed for a group G unless we assume some nice conditions on (G, ρ). In §4, we investigate in detail the Conway group $\cdot O$ and the Frame shapes with respect to the representation obtained from its action on the Leech lattice. As for its application, studies in moonshine have been made for $G = \mathrm{PSL}_2(\mathbb{F}_p)$ in [Kon1] and [Koi4].

§4. Moonshine for other groups (I)

Two ways of expressing Thompson series are shown in [CN], indicating two possibilities of constructing moonshine. These are discussed in §§4 and 5.

L is said to be an even lattice of rank d if L is a free $\mathbb{Z}$-module of rank d and if a positive definite quadratic form $\langle\ ,\ \rangle$ is defined on L satisfying $\langle x, x \rangle \in 2\mathbb{Z}$ and $\langle x, y \rangle \in \mathbb{Z}$ for all $x, y \in L$.

The theta function

$$\theta(z; L) = \sum_{x \in L} e^{\pi i \langle x, x \rangle z}$$

of the quadratic form associated with an even lattice L is an automorphic form of weight $\frac{d}{2}$.[9] We assume, furthermore, that a finite group G acts on L leaving $\langle\ ,\ \rangle$ invariant. We then define for an element g of G two types of automorphic forms $\theta_g(z)$ and $\eta_g(z)$ as follows:

(1) $L^g = \{x \in L \mid gx = x\}$ is a sublattice of L and, indeed, is an even lattice under the restriction of the quadratic form associated with L. Therefore, let $\theta_g(z)$ be the theta function of L^g:

$$\theta_g(z) = \theta(z; L^g)$$
$$= 1 + \sum_{n \geq 1} a_n(g) q^n;$$
$$a_n(g) = \#\{x \in L^g \mid \langle x, x \rangle = 2n\}.$$

(2) Let $\eta_g(z)$ be the η-product of the Frame shape of g with respect to the rational representation of G obtained from the action of G on L:

$$\eta_g(z) = q^{d/24} \sum_{n \geq 0} b_n(g) q^n.$$

Since $\eta_g(z)$ and $\theta_g(z)$ are then of equal weight, on taking their quotient we obtain an automorphic function

$$j_g(z) = \frac{\theta_g(z)}{\eta_g(z)}$$
$$= q^{-d/24} \sum_{n \geq 0} c_n(g) q^n.$$

---

[9] When d is odd, the form θ is called an automorphic form of half-integer weight, but we do not need this in our discussion of moonshine for now.

From the construction, these coefficients $a_n(g), b_n(g)$, and $c_n(g)$ are determined by the conjugacy class $\langle g \rangle$; indeed, in each case a representation space of G can be easily constructed to prove the following proposition.

PROPOSITION. (i) *For each $n \geq 1$, $a_n\colon g \mapsto a_n(g)$ is a character of G; for each $n \geq 0$, $c_n\colon g \mapsto c_n(g)$ is a character of G.*
(ii) *For each $n \geq 0$, $b_n\colon g \mapsto b_n(g)$ is a generalized character of G.*

It follows that when $d = 24$, $j_g(z)$ has a Fourier expansion of the same form as a Thompson series for the Monster; moreover, $j_g(z)$ satisfies the condition (M2).

Among lattices of rank 24, there is the Leech lattice Λ that is also interesting in connection with sphere packing and coding theory. As a characterization of Λ, we cite that Λ is an even unimodular lattice of rank 24 having no vector of length 2. In the language of automorphic forms, this means that the theta function of Λ belongs to $M_{12}(\Gamma_0(1), \mathrm{id})$[10] and the Fourier coefficient of q is 0.

The automorphism group of Λ is called the Conway group $\cdot O$, whose quotient group obtained when dividing by the center $\{\pm 1\}$ is a sporadic simple group. The group $\cdot O$ occupies an important position in the study of simple groups.[11] For the Leech lattice Λ (and $\cdot O$), $\theta_g(z)$ and $\eta_g(z)$ were completely determined by Kondo, Tasaka, Mason, Lang, and Koike. It may be said that the study has evolved into new riddles in connection with moonshine.

Let us write each element of $\cdot O$ in terms of its Frame shape with respect to the representation of $\cdot O$ obtained from the action of $\cdot O$ on Λ. Looking at the table[12] of Frame shapes, we can classify the elements $\pi = \prod t^{r_t}$ of $\cdot O$ into the following three kinds:

(i) π is of type C, which means that $r_t \geq 0$ for all t;
(ii) π is of type E, which means that $\mathrm{wt}\,\pi > 0$ but $r_t < 0$ for some t;
(iii) π is of type F, which means $\mathrm{wt}\,\pi = 0$.
We also note that $\mathrm{wt}\,\pi \in \mathbb{Z}$ for all π.

THEOREM 2.
(1) *If π is of type C, then $\eta_\pi(z)$ is a cusp form and is a common eigenfunction for all Hecke operators.*
(2) *If π is of type E, then $\eta_x(z)$ is an Eisenstein series possessing multiplicative Fourier coefficients and is, futhermore, an eigenfunction for all Hecke operators T_p except possibly for one prime p.*
(3) *If π is of type F, then $\eta_\pi(z)$ is an automorphic function and there exists an element m of M such that*

$$\eta_\pi(z) = t_m(z)^{-1}.$$ [13]

---

[10] The weight 12 is also the smallest k satisfying $S_k(\Gamma_0(1), \mathrm{id}) \neq 0$.

[11] Definitions of the Leech lattice and its relations to simple groups such as M_{24} are comprehensibly explained in [C3].

[12] There is a table of Frame shapes of $\cdot O$ in [Kon1]. It seems necessary to have the table in order to understand the formulas for the Thompson series in Table 3 of [CN].

[13] It was necessary to make the constant term 0 in the definition of a Thompson series because it is impossible to make a character of M there. However, as the later Fourier coefficients will not change by adding a constant term, it is convenient to use the notation $t_m(z) = T_m(z) +$ constant, and the constant term will not be written whenever this is clear from the context. For example, in the problem that follows, the constant term in $\theta_\pi(z)/\eta_\pi(z)$ can be readily calculated from $\eta_\pi(z)$, since Λ does not have a vector of length 2.

The fact that π of type C satisfies condition (B) of Theorem 1 yields a proof of (1). Since all the functions considered in (2) and (3) can be written explicitly, comparison of the Fourier coefficients yields the results.

The Frame shapes can be calculated when the character table and the power products are known, and their η-products can also be investigated. However, the determination of $\theta_g(z)$ is difficult, and even the description of the results is not as simple as in the above theorem. To illustrate the difficulty, let us state a theorem in connection with the following problem that was stated in [CN] in relation to moonshine.

PROBLEM. For each element π of $\cdot O$, determine whether there exists some element m of M satisfying

(♮)
$$\frac{\theta_\pi(z)}{\eta_\pi(z)} = t_m(z).\text{[14]}$$

If the answer is affirmative, then moonshine for $\cdot O$ can be constructed through Λ and the Thompson series of M can also be studied from Λ.[15] Concerning this problem, the following definitive result has been obtained by Kondo–Tasaka [KT1, 2], Lang [L], and Koike [Koi5].

THEOREM 3. *For each element π of $\cdot O$, except for those in one of 15 exceptional conjugacy classes, there exists an element m of M such that (♮) holds. Every exceptional conjugacy class is of type E.*

Even for any exceptional conjugacy class $\langle \pi \rangle$, the left-hand side of (♮) can be expressed as a sum and difference of three Thompson series. It would be interesting to find an explanation of why such exceptions occur.

Interesting facts have been observed about $\theta_\pi(z)$ itself, too, in connection with the action of the Atkin-Lehner involutions. See [Koi5].

§5. Moonshine for other groups (II)

Let p be a prime such that $(p-1)|24$, i.e., $p = 2, 3, 5, 7$, or 13. Put $2d = 24/(p-1)$. There exists an element π_p of $\cdot O$ of order p having no fixed point in Λ. The centralizer of π_p in $\cdot O$ is written below:

p	the centralizer of π_p
2	$\cdot O$
3	$6S_Z$
5	$5 \times 2HJ$
7	$7 \times 2A_7$
13	$13 \times \mathrm{SL}_2(3)$.

---

[14] This formula can be regarded as one that determines a theta function of some quadratic form satisfying (♮) when $\eta_\pi(z)$ is given. The uniqueness and existence of $\theta_\pi(z)$ can be investigated from this standpoint. See [Koi3, 4].

[15] Indeed, Λ and the moonshine module in [FLM] are related.

See [**Atlas**] for the notation to describe the centralizers. Each of $\cdot O, 6S_Z, 2HJ$, $2A_7$, and $SL_2(3)$ has a $2d$-dimensional representation. Although some of these representations are not rational representations over $\mathbb{Q}$, Frame shapes can be extended and defined. Then for each p, when the Frame shape of a group element can be written $\pi = \prod t^{r_t}$, consider

$$\pi \circ (p) = \prod t^{r_t} \cdot (pt)^{-r_t}.$$

The weight of $\pi \circ (p)$ is 0, and the degree is -24. It follows that the corresponding η-product has the same form as a Thompson series and the Fourier coefficients form generalized characters in the light of the proposition in §4.

In the case $p = 2$, the representation is rational over $\mathbb{Q}$ and $\eta_{\pi \circ (2)}$ is always the type belonging to a group of genus 0. It is called multiplicative moonshine in [**CN**]. In [**Q**], $\eta_{\pi \circ (p)}(z)$ is given for other p also, but automorphic functions with Fourier coefficients, which are not rational, also arise and their properties are unclear.

§6. Replication formulas

A sequence of formulas, called replication formulas, exhibits some of the most interesting properties satisfied by the Thompson series. These properties are similar to those of Hecke operators. Considering that Hecke operators did not play as important a role in automorphic function theory as in automorphic form theory and that these formulas are valid only for automorphic functions of genus 0, we can understand why these formulas have not been found until recently. It seems that these formulas could be discovered only by recognizing the fact in the background that the group structure of M is reflected beautifully in the Thompson series.

A formal power series of the form

$$t = q^{-1} + \sum_{n \geq 0} a_n q^n, \qquad a_n \in \mathbb{C}$$

is called a type. For each type t, a polynomial $X(n, t)$ of degree n in t can be uniquely determined satisfying

$$X(n, t) \equiv \frac{1}{n} q^{-n} \pmod{\mathscr{R}}, \qquad \mathscr{R} = q\mathbb{C}[[q]].$$

DEFINITION. Let p be a prime. If there exists a type $t^{(p)}$ satisfying

$$(*) \qquad X(n, t)|U_p + X(n, t^{(p)})|V_p = pX(np, t) + X\left(\frac{n}{p}, t^{(p)}\right)$$

for all $n \geq 1$, then $t^{(p)}$ is called a p-plicate of t. Here U_p and V_p are those defined in §1, and it is to be understood that $X(n/p, t^{(p)}) = 0$ when $p \nmid n$.

If we compare the left-hand side of the formula $(*)$ with the Hecke operator defined in §1 with weight $k = 0$, we notice the only difference is that, within the part where V_p operates, t is replaced by $t^{(p)}$.

This definition is abstract and, although we can seek conditions on t for t to have a p-plicate independent of automorphicity, we cannot say anything yet in general.[16] From $(*)$ for $n = 1$, we see that $t^{(p)}$ is unique whenever it exists.

---

[16] In [**N**], the definition, the problem, and the conjecture are stated slightly differently. The appendix by Norton in [**M3**] is also interesting.

The formula (∗) was discovered by Conway and Norton when they investigated the action of Hecke operators on the Thompson series using the group structure of M, but the formula can be handled slightly more generally within the framework of automorphic functions. However, to keep the presentation from becoming too complicated, we restrict ourselves to a simple case.

Let $\Gamma = N + S$ be of genus 0. For each prime p, let

$$N^{(p)} = N/(p, N),$$
$$S^{(p)} = \{Q \in S \mid Q \text{ is a divisor of } N^{(p)}\},$$

and consider $\Gamma^{(p)} = N^{(p)} + S^{(p)}$.

It can then be ascertained from the table[17] of groups of genus 0 of the form $\Gamma = N + S$ that $\Gamma^{(p)}$ is also of genus 0.

THEOREM 4. *With the notation above, the type belonging to $\Gamma^{(p)}$ is the p-plicate of the type belonging to Γ.*

We prove this using the following two lemmas, case-by-case, according to how N, S, and p are related. To avoid technicalities, we write the result for the simplest case.

Case. $p \| N$ and $p \in S$.

Let t and $t^{(p)}$ be the types belonging to Γ and $\Gamma^{(p)}$, respectively.

LEMMA 1. $X(n, t) | U_p + X(n, t) = X(n, t^{(p)}) + X(n/p, t^{(p)})$.

LEMMA 2. $X(n, t^{(p)}) + X(n, t^{(p)}) | V_p = X(n, t) + p X(pn, t)$.

To prove Lemma 1, observe that for the automorphic function $X(n, t)$ with respect to Γ, the left-hand side of the equality becomes an automorphic function with respect to $\Gamma^{(p)}$ and, since the genus is 0, the right-hand side is obtained by checking where the poles are. To prove Lemma 2, investigate in the reverse direction.

It is clear that formula (∗) follows from Lemmas 1 and 2.

Write $n = p_1 p_2 \cdots p_s$, a product of primes. Starting from t, let t_1 be the p_1-plicate of $t, \ldots$, let t_i be the p_i-plicate of t_{i-1}, and so on. We can determine $t_1, t_2, \ldots, t_s$ in succession and, from the procedure of construction of $\Gamma^{(p)}$, t_s is independent of the ordering of $p_1, p_2, \ldots, p_s$. Hence, t_s can be defined as the n-plicate of t, for which we write $t^{(n)}$.

THEOREM 5. *For an element g of M, let g^p be the pth power of g. Then the p-plicate of the Thompson series t_g is t_{g^p}.* [18]

Finally, we define Hecke operators. For a power p^a of a prime p, we define the Hecke operator T_{p^a} by

$$X(n, t) | T_{p^a} = \sum_{i=0}^{a} p^{-a} X(n, t^{(p^i)}) | U_p^{a-i} | V_p^i.$$

By using the same argument as in the classical case, we see that the T_{p^a}'s are mutually

---

[17] There is such a table in [H].

[18] Raising an element m of M to the nth power and forming the n-plicate of the corresponding Thompson series $T_m(z)$ are compatible. The moonshine for M_{24} constructed by using the Leech lattice Λ in §4 satisfies (M1)′ and (M2)′, but does not satisfy the property corresponding to Theorem 5. Will not a moonshine for M be uniquely determined if Theorem 5 is added to (M0), (M1), and (M2)?

commutative, and the following formula $(**)$ can be obtained as a generalization of the formula $(*)$:

$$(**) \qquad X(n,t)|T_{p^a} = \sum_{p^j|(n,p^a)} p^{-j} X(p^{a-2j}n, t^{(p^j)}).$$

Then we can consider the commutative ring consisting of all Hecke operators. The relation formulas, assuming that p and ℓ are distinct primes, can be written as follows:

(i) $T_{p^a} \circ T_{\ell^b} = T_{\ell^b} \circ T_{p^a}$,
(ii) $T_{p^a} \circ T_p = T_{p^{a+1}} + (1/p)I_p \circ T_{p^{a-1}}$, $a \geq 1$, where $X(n,t)|I_p = X(n, t^{(p)})$,
(iii) $I_p \circ I_\ell = I_\ell \circ I_p$.

§7. Concluding remarks

So far, nothing has been written here about the group-theoretic properties of the Monster owing to ignorance on the part of the author. While its construction was simplified by Conway and Tits, following the proof by Griess, progress was made in connection with an attempt towards a proof of the moonshine conjecture, leading to the construction of a graded module by Frenkel, Leopowsky, and Meurman on which the Monster acts (called a moonshine module). But here again, the appearance of "automorphic functions of genus 0" could not be explained.

When will the moonshine be broken through by the sunlight?

Acknowledgment

The translator is grateful to Professors J. F. Glazebrook, P. R. Hewitt, and Michio Suzuki for their invaluable advice. Professor Suzuki's influence is evident throughout.

References

[C1] J. H. Conway, *Monster and Moonshine*, Math. Intelligencer **2** (1980), 165–171.
[C2] _____, *A simple construction for the Fischer-Griess monster group*, Invent. Math. **79** (1985), 513–540.
[C3] _____, *Three lectures on exceptional groups, finite simple groups*, M. Powell and G. Higman, eds., Academic Press, New York, 1971.
[CN] J. H. Conway and S. P. Norton, *Monstrous moonshine*, Bull. London Math. Soc. **11** (1979), 308–339.
[DKM] D. Dummit, and J. Mckay, H. Kisilevsky, *Multiplicative η-products*, Contemp. Math., vol. 45, Amer. Math. Soc., Providence, R.I., 1985, pp. 89–98.
[FLM] I. Frenkel, J. Lepowsky, and A. Meurman, *A moonshine module for the Monster*, Vertex operators in Mathematics and Physics, Publ. Res. Inst. Math. Sci. **3** (1984), 231–273.
[G] R. L. Griess, *The friendly giant*, Invent. Math. **69** (1982), 1–102.
[H] K. Harada, *"Moonshine" of finite groups*, Lecture Notes, The Ohio State University, Columbus, Ohio.
[Koi1] M. Koike, *On McKay's conjecture*, Nagoya Math. J. **95** (1984), 85–89.
[Koi2] _____, *Moonshine for* $PSL_2(F_7)$, Adv. Stud. Pure Math. **7** (1985), 103–111.
[Koi3] _____, *Mathieu group M_{24} and modular forms*, Nagoya Math. J. **99** (1985), 147–157.
[Koi4] _____, *Moonshines of $PSL_2(F_q)$ and the automorphism group of Leech lattice*, Japan. J. Math. **12** (1986), 283–323.
[Koi5] _____, *Modular forms and the automorphism group of Leech lattice*, Nagoya Math. J. **112** (1988), 63–79.
[Koi6] _____, *On replication formula and Hecke operators*, preprint.
[Kon1] T. Kondo, *The automorphism group of the Leech lattice and elliptic modular functions*, J. Math. Soc. Japan **37** (1985), 337–361.

[Kon2] _____, *Examples of multiplicative η-products*, Sci. Papers College Arts Sci. Univ. Tokyo **35** (1985), 133–149.
[KT1] T. Kondo and T. Tasaka, *The theta functions of sublattices of the Leech lattice*, Nagoya Math. J. **101** (1985), 151–179.
[KT2] _____, *The theta functions of sublattices of the Leech lattice*. II, J. Fac. Sci. Univ. Tokyo **34** (1987), 545–572.
[L] M. L. Lang, *On a question raised by Conway-Norton*, J. Math. Soc. Japan **41** (1989), 263–284.
[M1] G. Mason, *Modular forms and the theory of Thompson series*, Proc. Rutgers Group Theory year 1983–1984 (Aschbacher, ed.), Cambridge Univ. Press, London and New York, 1984, pp. 391–407.
[M2] _____, M_{24} *and certain automorphic forms*, Contemp. Math., vol. 45, Amer. Math. Soc., Providence, R.I., 1985, pp. 223–244.
[M3] _____, *Finite groups and modular functions*, Proc. Sympos. Pure Math., vol. **47**, Amer. Math. Soc., Providence, R.I., 1987, pp. 181–210.
[M4] _____, *Finite groups and Hecke operators*, Math. Ann. **283** (1989), 381–409.
[N] S. P. Norton, *More on moonshine*, Computational Group Theory, Academic Press, New York, 1984, pp. 185–193.
[Q] L. Queen, *Modular functions arising from some finite groups*, Math. Comp. 37 (1981), 547–580.
[S] S. F. Smith, *On the Head characters of the Monster simple group*, Contemp. Math., Vol. **45**, Amer. Math. Soc., Providence, R.I., 1985, pp. 303–313.
[Atlas] J. H. Conway et al., *Atlas of finite groups*, Clarendon Press, Oxford, Amer. Math. Soc., Providence, R.I., 1985.

DEPARTMENT OF MATHEMATICS, FACULTY OF SCIENCE, HIROSHIMA UNIVERSITY, HIGASHI-HIROSHIMA, JAPAN

Translated by N. C. HSU

On the Theory of Mixed Hodge Modules

Morihiko Saito

Recently Beilinson, Bernstein, Deligne, and Gabber [3] made further progress in the theory on the Weil conjecture in characteristic p. By analogy between two theories [12], it stimulated some progress in the Hodge theory: e.g., the existence of a pure Hodge structure on the intersection cohomology with coefficients in a polarizable variation of Hodge structure defined on the complement of a normal crossing divisor by Cattani-Kaplan-Schmid [10] and Kashiwara-Kawai [30, 31], and also the theory of admissible variation of mixed Hodge structure by Steenbrink-Zucker [60] and Kashiwara [27], etc. Further developments were expected with the introduction of new objects that correspond philosophically to mixed perverse sheaves in characteristic p [3]. This hope was recently realized as the theory of mixed Hodge modules [39, 40], by using the theory of filtered $\mathscr{D}$-modules and vanishing cycles. We explain this development. See also [47, 54] for other introductions to the theory of mixed Hodge modules.

In §1, we review Hodge theory. We state the main theorems in §2, after explaining the analogy with the Weil conjecture. We review some elementary facts from the theory of filtered $\mathscr{D}$-modules in §3 and explain the idea of the proof of the main theorem in §4. Some applications are mentioned in §5.

In this paper, algebraic varieties are defined over $\mathbb{C}$ and assumed separated.

§1. Hodge theory

Let X be a compact Kähler manifold. By Hodge theory, the cohomology groups of X have the Hodge decomposition

$$(1.1) \qquad H^n(X, \mathbb{C}) = \bigoplus_{p+q=n} H^{p,q},$$

where $H^{p,q}$ denotes the space of harmonic forms of type (p, q). This decomposition appears to depend on the choice of Kähler form on X. But we can show the

---

1991 *Mathematics Subject Classification.* Primary 14C30.

This article originally appeared in Japanese in Sūgaku **43** (4) (1991), 289–301.

uniqueness of the decomposition as follows. We define the Hodge filtration F by $F^i = \bigoplus_{p \geq i} H^{p,q}$ so that

(1.2) $$H^{p,q} = F^p \cap \overline{F^q},$$

where $\overline{F^q}$ is the complex conjugate of F^q. Then F^p is expressed as

(1.3) $$F^p = \mathbf{H}^n(X, \sigma_{\geq p}\Omega_X^{\cdot}),$$

using holomorphic differential forms $\Omega_X^{\cdot}$ (see [12]). Here $\sigma_{\geq p}\Omega_X^{\cdot}$ is the subcomplex of the de Rham complex $\Omega_X^{\cdot}$, whose ith component is the same as Ω_X^i for $i \geq p$, and 0 otherwise.

In general, we say that a finite decreasing filtration F on $H_{\mathbb{C}} := H \otimes_{\mathbb{Q}} \mathbb{C}$ gives a *Hodge structure of weight* n on a finite-dimensional vector space H over $\mathbb{Q}$ if the properties like (1.1)–(1.2) are satisfied. A Hodge structure (H, F) is called *polarizable* if there exists a bilinear form S on H such that $S(F^p, F^{n-p+1}) = 0$, $S(u, v) = (-1)^n S(v, u)$, and $i^{q-p} S(u, \overline{u}) > 0$ for a nonzero $u \in H^{p,q}$. Here S is called a *polarization* of the Hodge structure. In the above case, we have a polarization of the Hodge structure on the cohomology of X by means of a Kähler form.

Let X be an algebraic variety. By Deligne [12], the cohomology of X has a polarizable mixed Hodge structure. In general, we say that a finite-dimensional vector space H over $\mathbb{Q}$ has a *mixed Hodge structure* if there are a finite increasing filtration W on H and a finite decreasing filtration F on $H_{\mathbb{C}}$ such that F gives a Hodge structure of weight n on $\mathrm{Gr}_n^W H := W_n H / W_{n-1} H$. Here F and W are called the *Hodge* and *weight filtrations* respectively. We say that a mixed Hodge structure is *polarizable* if each graded piece $\mathrm{Gr}_n^W H$ is a polarizable Hodge structure.

Let $f : X \to S$ be a proper smooth morphism of smooth algebraic varieties, and L the higher direct image $R^n f_* \mathbb{Q}_X$. Since f induces a topological fibration over S, L is a local system whose stalk is the cohomology of the fiber of f. Let $\mathscr{L}$ denote the corresponding analytic vector bundle (i.e., locally free sheaf) $L \otimes \mathscr{O}_S$. Then the Hodge filtration F on the cohomology of each fiber determines a filtration F of $\mathscr{L}$ by analytic subbundles, and we have *Griffiths' transversality* [20]:

(1.4) $$\xi F^p \mathscr{L} \subset F^{p-1} \mathscr{L}$$

for a holomorphic vector field ξ. In fact, let $\Omega_{X/S}^{\cdot}$ be the sheaf complex of relative differential forms. Then $\mathscr{L}$ is naturally isomorphic to $R^n f_* \Omega_{X/S}^{\cdot}$, and the Hodge filtration F on $\mathscr{L}$ is expressed by using the truncation σ as in (1.3). Furthermore, (1.4) is verified by using the Gauss-Manin connection (or filtered $\mathscr{D}$-modules).

In general, we say that (L, F) is a *variation of Hodge structure of weight* n if L is a local system with $\mathbb{Q}$-coefficients on a complex manifold S and F is a filtration of $\mathscr{L} := L \otimes \mathscr{O}_S$ by analytic subbundles such that F determines a Hodge structure of weight n at each fiber of L and (1.4) is satisfied. A *polarization* of a variation of Hodge structure is a pairing of local systems $L \otimes L \to \mathbb{Q}_S$ that induces a polarization of Hodge structure at each point of S. In the above geometric case, we have a polarization if X is Kähler or f is projective.

If S is compact Kähler, the cohomology with coefficients in a polarizable variation of Hodge structure on S has a canonical Hodge structure [65]. Furthermore, its

Hodge filtration F is given by taking the hypercohomology of the Hodge filtration F of $\Omega_S^{\cdot} \otimes \mathscr{L}$ that is defined by

(1.5) $$F^p = [F^p\mathscr{L} \to \cdots \to \Omega_S^d \otimes F^{p-d}\mathscr{L}],$$

where $d = \dim S$. Note that this is a generalization of the constant coefficient case (1.3).

However, polarizable variations of Hodge structures defined on compact manifolds are not so interesting in some cases. For example, they must be constant if $S = \mathbb{P}^1$. So we are interested in how Hodge structures degenerate at the boundary of the manifold in the noncompact case. This is related to the problem of the degeneration of a family of algebraic varieties in the case where the variation is defined geometrically as above. These problems were studied by Schmid [56] and Steenbrink [57] in the one-dimensional case.

Let S be an open disc with coordinate t, and $S^* = S \setminus \{0\}$. Let (L, F) be a polarizable variation of Hodge structure of weight n on S^* whose monodromy T is unipotent. By [13], $\mathscr{L} := L \otimes \mathscr{O}_{S^*}$ is naturally extended to an analytic vector bundle $\hat{L}$ with a regular singular connection. Let $\hat{L}(0) = \hat{L}/t\hat{L}$ (the fiber of $\hat{L}$ at 0), and L_∞ the multivalued sections of L. Then we have an isomorphism

(1.6) $$\hat{L}(0) = L_\infty \otimes \mathbb{C}$$

by [13]. (Note that $L_\infty = \psi_t L$, and (1.6) is a special case of (3.7) below, because the filtration V on $\hat{L}[t^{-1}]$ is given by $t^i\hat{L}$.) By Schmid's nilpotent orbit theorem [56], the Hodge filtration F on $\mathscr{L}$ is naturally extended to the Hodge filtration F of $\hat{L}$ by analytic subbundles, and it determines a filtration F on $L_\infty \otimes \mathbb{C}$ by (1.6). Furthermore, we have that

(1.7) $\qquad (L_\infty, F, W)$ is a mixed Hodge structure,

as a corollary of the SL_2-orbit theorem. See [56]. Here W is the monodromy filtration associated with the action of $N = \log T$ up to the shift by n; i.e., it is determined by the properties: $NW_i \subset W_{i-2}, N^j : \operatorname{Gr}_{n+j}^W \xrightarrow{\sim} \operatorname{Gr}_{n-j}^W (j > 0)$. Note that (1.7) is the simplest case of the stability by ψ in Theorem 2.1 below.

Now let $f : X \to S$ be a projective morphism of a complex manifold X onto S (i.e., X may be assumed a closed submanifold of $\mathbb{P}^m \times S$). Let $f' : X^* \to S^*$ be the restriction of f over S^*. Assume f' smooth and $Y := f^{-1}(0)$ a reduced divisor with normal crossings. In the case where the above (L, F) is given by the variation of Hodge structure on the higher direct image $R^n f'_* \mathbb{Q}_{X^*}$, a geometric proof of (1.7) was obtained by Steenbrink [57].

He constructed a weight filtration W on Deligne's nearby cycle sheaf complex $\psi_f \mathbb{Q}_X$ [14], which is nothing but the monodromy filtration defined in the abelian category of perverse sheaves by a recent terminology [3]. Furthermore, its primitive part $P\operatorname{Gr}_i^W := \operatorname{Ker} N^{i+1} \subset \operatorname{Gr}_i^W$ is a direct sum of the constant sheaves supported on the intersections of $i + 1$ irreducible components of Y and shifted by i to the right. This decomposition is compatible with the Hodge filtration, and we get a proof of (1.7) by showing the commutativity of the functor ψ with the direct image using Hodge theory. A key point is the stability of the monodromy filtration by the direct image, which is not so trivial because we have to treat a weight spectral sequence not degenerating at E_1 (see [39, 4.2.5]).

To generalize the above argument to the higher-dimensional case, it is better to use filtered $\mathscr{D}$-modules instead of vector bundles with a filtration and a connection.

In fact, combining and further developing the theory of filtered $\mathscr{D}$-modules and the theory of nearby and vanishing cycles [14, 28, 34], we can get the theory of mixed Hodge modules. In the next section, we first explain the relation with the Weil conjecture as the philosophical background which made possible such a theory.

§2. Hodge module

By Deligne's dictionary [12], there is some analogy between the theory of the Weil conjecture concerning the absolute value of the eigenvalues of the Frobenius operator on a variety of characteristic p and the theory of mixed Hodge structure on a complex algebraic variety. A key point is the notion of weight, which is well formulated in terms of weight filtration. These are the so-called absolute theories (i.e., concerning the cohomology of varieties). In characteristic p, the relative version of the Weil conjecture was obtained by Deligne as the theory of mixed complexes [15]. Combined with the theory of t-structure, it developed further to become the theory of mixed perverse sheaves by Beilinson, Bernstein, Deligne, and Gabber [3]. So, by generalizing the above analogy, it has been conjectured that there would exist objects in characteristic 0 that correspond philosophically to mixed perverse sheaves in characteristic p and contain polarizable variations of Hodge structures as special cases. For example, the conjecture on a pure Hodge structure on intersection cohomology is considered as strong evidence to the above conjecture. These conjectures are all positively solved by introducing the notion of mixed Hodge modules. Here the analogy between the mixed Hodge modules and the mixed perverse sheaves means that both satisfy the formalism of mixed sheaves. For the mixed Hodge modules, it is expressed by the following theorem and proposition (cf. [40]) :

THEOREM 2.1. *For an algebraic variety X, there exists an abelian category* $\mathrm{MHM}(X)$ *consisting of mixed Hodge modules on X, together with a faithful and exact functor* $\mathrm{rat} : \mathrm{MHM}(X) \to \mathrm{Perv}(\mathbb{Q}_X)$, *and we have naturally standard functors* f_*, $f_!$, f^*, $f^!$, ψ_g, φ_g, $\mathbb{D}$, $\boxtimes$, $\otimes$, $\mathscr{H}om$ *between the bounded derived categories* $D^b \mathrm{MHM}(X)$, *so that these functors commute with the composition*

(2.1) $\qquad \mathrm{rat} : D^b \mathrm{MHM}(X) \to D^b \mathrm{Perv}(\mathbb{Q}_X) \simeq D^b_c(\mathbb{Q}_X).$

REMARK. In (2.1), $D^b_c(\mathbb{Q}_X)$ denotes the derived category of bounded complexes of $\mathbb{Q}_X$-modules such that the restriction of their cohomology to each stratum of an algebraic stratification is a local system, and $\mathrm{Perv}(\mathbb{Q}_X)$ is its full subcategory consisting of perverse sheaves [3] defined by the condition (3.2) below. The equivalence of categories in (2.1) is due to [2, 3]. For $\mathscr{M} \in D^b \mathrm{MHM}(X)$, $\mathrm{rat}(\mathscr{M})$ is called the *underlying $\mathbb{Q}$-complex* of $\mathscr{M}$.

In $D^b_c(\mathbb{Q}_X)$, the above functors are constructed by Grothendieck, Verdier [63], and Deligne [14]. The above theorem means that these functors can be naturally lifted to the functors between the bounded derived categories of mixed Hodge modules.

PROPOSITION 2.2. *Each mixed Hodge module $\mathscr{M}$ has a functorially defined finite increasing filtration W (called the weight filtration) such that $\mathscr{M} \to \mathrm{Gr}^W_i \mathscr{M}$ is an exact functor, and $\mathrm{Gr}^W_i \mathscr{M}$ are semisimple.*

REMARK. The axioms of mixed sheaves [51] include the existence of constant sheaves on smooth varieties. In the case of mixed Hodge modules, it follows from

Theorem 2.1 and the following equivalence of categories, by using the vector space $\mathbb{Q}$ with a trivial Hodge structure of type $(0,0)$. (Here pt = Spec $\mathbb{C}$.)

THEOREM 2.3. MHM(pt) $\simeq$ {*polarizable mixed Hodge structures*}.

REMARK. In the case where the constant sheaf exists, we can construct the functors in Theorem 2.1 by the general theory of mixed sheaves [51], if we have the cohomological direct image $H^j f_*$ for an affine morphism f, the pull-back j^* for an embedding j, the dual functor $\mathbb{D}$, and the external product $\boxtimes$, so that they satisfy some natural properties associated with these functors. In the case of mixed Hodge modules, the constant sheaf on a smooth variety X is given by the constant sheaf $\mathbb{Q}_X$ with a trivial variation of Hodge structure of type (0,0). However, note that the functors ψ_g, φ_g are used in an essential way to construct the direct image $H^j f_*$ in this case.

In the following, we identify both sides of Theorem 2.3 and denote by $\mathbb{Q}^H$ the trivial Hodge structure $\mathbb{Q}$ of type $(0,0)$. For an algebraic variety X, let $a_X : X \to$ pt be a natural morphism, and $\mathbb{Q}_X^H = a_X^* \mathbb{Q}^H \in D^b \mathrm{MHM}(X)$. Then

$$(2.2) \qquad \mathrm{rat}(a_X)_* \mathbb{Q}_X^H = (a_X)_* \mathbb{Q}_X \ (= \mathbf{R}\Gamma(X, \mathbb{Q})),$$

and we get a natural mixed Hodge structure on the cohomology of X. A similar argument holds for cohomology with compact supports, local cohomology, and (Borel-Moore) homology, if we use $(a_X)_!$, etc. We can check easily the coincidence with Deligne's mixed Hodge structure [12] in the case where X is a subvariety of a smooth variety (e.g., X quasiprojective).

DEFINITION 2.4. We say that $\mathscr{M} \in D^b \mathrm{MHM}(X)$ has *weights* $\leq n$ (resp. $\geq n$) if $\mathrm{Gr}_j^W H^i \mathscr{M} = 0$ for $j > n+i$ (resp. $j < n+i$). If $\mathscr{M}$ has weights $\leq n$ and weights $\geq n$, $\mathscr{M}$ is called *pure of weight n*.

PROPOSITION 2.5. *The condition weights $\leq n$ (resp. $\geq n$) is stable by $f_!, f^*$ (resp. $f_*, f^!$). In particular, pure Hodge complexes are stable by the direct image under a proper morphism.*

PROPOSITION 2.6. *If $\mathscr{M} \in D^b \mathrm{MHM}(X)$ is pure, we have the following (noncanonical) isomorphism in $D^b \mathrm{MHM}(X)$:*

$$(2.3) \qquad \mathscr{M} = \bigoplus_i (H^i \mathscr{M})[-i].$$

This follows from the semisimplicity of pure Hodge modules. From Propositions 2.5 and 2.6, we can deduce the Beilinson-Bernstein-Deligne-Gabber type decomposition theorem.

DEFINITION 2.7. Let $\mathrm{MH}(X,n)^p$ be the full subcategory of $\mathrm{MHM}(X)$ consisting of pure Hodge modules of weight n. Let Z be a closed irreducible subvariety of X. We say that $\mathscr{M}$ has *strict support* Z if the support of $\mathscr{M}$ is Z and $\mathscr{M}$ has no nontrivial sub or quotient object with smaller support. Let $\mathrm{MH}_Z(X,n)^p$ be the full subcategory of $\mathrm{MH}(X,n)^p$ consisting of pure Hodge modules with strict support Z or $\emptyset$.

Note that $\mathrm{MH}(X,n)^p$ is a semisimple abelian category and its object has the canonical decomposition:

$$\tag{2.4} \mathscr{M} = \bigoplus_Z \mathscr{M}_Z \quad \text{with} \quad \mathscr{M}_Z \in \mathrm{MH}_Z(X,n)^p.$$

In particular, any mixed Hodge module is obtained by successive extensions of pure Hodge modules with strict supports.

REMARK. A pure Hodge module is called a *polarizable Hodge module* in [39], and the notation $\mathrm{MH}(X,n)^p$ comes from this.

THEOREM 2.8. $\mathrm{MH}_Z(X,n)^p \simeq \mathrm{VHS}_{\mathrm{gen}}(Z, n - \dim Z)^p$.

Here the right-hand side is the category of polarizable variations of Hodge structures of weight $n - \dim Z$ defined on nonempty smooth open subvarieties of Z. This theorem means that the restriction of any pure Hodge module to a sufficient small open subvariety is a polarizable variation of Hodge structure, and conversely, any polarizable variation of Hodge structure on a smooth open subvariety is extended uniquely to a pure Hodge module with strict support. In particular, $\mathrm{MH}_Z(X,n)^p$ depends only on Z.

DEFINITION 2.9. Let $\mathscr{M} = (L, F)$ be a polarizable variation of Hodge structure on a smooth open variety U of Z. We denote by $\mathrm{IC}_Z \mathscr{M}$ the pure Hodge module with strict support Z, which corresponds to $\mathscr{M}$ by 2.8.

REMARK. The polarizable variation of Hodge structure $\mathscr{M}$ is identified with a pure Hodge module on U, and we have a canonical isomorphism

$$\tag{2.5} \mathrm{IC}_Z \mathscr{M} = j_{!*}\mathscr{M} := \mathrm{Im}(H^0 j_! \mathscr{M} \to H^0 j_* \mathscr{M}),$$

where $j : U \to Z$ denotes a natural inclusion. This implies $\mathrm{rat}(\mathrm{IC}_Z \mathscr{M}) = \mathrm{IC}_Z L$, where $\mathrm{IC}_Z L$ is defined by $j_{!*}(L[-\dim Z])$ as above and is called the intersection complex with coefficients in L (see [3]). Combining this with Theorem 2.1 and Proposition 2.5, we get a canonical pure Hodge structure on the intersection cohomology $\mathrm{IH}^{\cdot}(Z, L)$ in the Z compact case. This Hodge structure coincides with that of Cattani-Kaplan-Schmid [10] and Kashiwara-Kawai [30, 31] in the case where Z is smooth and $Z \setminus U$ is a divisor with normal crossings.

For mixed Hodge modules, we do not have the decomposition by strict support (2.4), and we have to consider all possible extensions rather than the extensions in (2.5). For a function g on X, let $Y = g^{-1}(0)$, and $U = X \setminus Y$. We denote by $\mathrm{MHM}(U, Y, g)$ the category consisting of $(\mathscr{M}', \mathscr{M}'', u, v)$, where $\mathscr{M}' \in \mathrm{MHM}(U)$, $\mathscr{M}'' \in \mathrm{MHM}(Y)$, $u : \psi_{g,1}\mathscr{M}' \to \mathscr{M}''$, $v : \mathscr{M}'' \to \psi_{g,1}\mathscr{M}'(-1)$ such that $vu = N$, and the morphisms are pairs of morphisms of $\mathscr{M}', \mathscr{M}''$ that commute with u, v. Here $\psi_{g,1}$ is the direct factor of ψ_g with unipotent monodromy.

PROPOSITION 2.10. $\mathrm{MHM}(X) \simeq \mathrm{MHM}(U, Y, g)$.

This means that mixed Hodge modules glue together using the nearby cycle functor. See [**40, 2.28**]. In the case of perverse sheaves, it was obtained by Verdier [64].

DEFINITION 2.11. We say that a mixed Hodge module $\mathscr{M}$ on a smooth variety X is *smooth* if $\mathrm{rat}(\mathscr{M})[-\dim X]$ is a local system on X. We denote by $\mathrm{MHM}(X)_s$ the full subcategory of smooth mixed Hodge modules on X.

THEOREM 2.12. $\mathrm{MHM}(X)_s = \mathrm{VMHS}(X)_{\mathrm{ad}}$.

Here the right-hand side is the category of admissible variations of mixed Hodge structures in the sense of Steenbrink-Zucker [60] and Kashiwara [27]. They are variations of mixed Hodge structures such that (1.4) is satisfied, Gr_i^W are polarizable variations of Hodge structures, and, furthermore, in the one-dimensional case with unipotent monodromy, the relative monodromy exists and the limit of Hodge filtration behaves well around the compactified points. In general, they are defined by reducing to the above case, restricting to curves, and taking ramified coverings.

By Proposition 2.10 and Theorem 2.12, mixed Hodge modules can be constructed by induction on dimension, gluing admissible variations of mixed Hodge structures. This is used in an essential way to prove the stability of mixed Hodge modules by external products. Note that the cohomology with coefficients in an admissible variation of mixed Hodge structure has a canonical mixed Hodge structure by Theorems 2.1 and 2.12, and it can be used to get the short exact sequence [43]:

$$0 \to \mathrm{Ext}^1(\mathbb{Q}^H, H^0(X, \mathscr{M}^* \otimes \mathscr{N})) \to \mathrm{Ext}^1(\mathscr{M}, \mathscr{N}) \to \mathrm{Hom}(\mathbb{Q}^H, H^1(X, \mathscr{M}^* \otimes \mathscr{N})) \to 0$$

for $\mathscr{M}, \mathscr{N} \in \mathrm{VMHS}(X)_{\mathrm{ad}}$, where $\mathscr{M}^*$ is the dual variation of $\mathscr{M}$. Applying this to $W^{n-1}\mathscr{M}$ and $\mathrm{Gr}_n^W \mathscr{M}$, the condition for admissible variation of mixed Hodge structure can be understood globally in some sense.

§3. $\mathscr{D}$-module

In this section, we review some elementary facts from the theory of filtered $\mathscr{D}$-modules which will be needed in the construction of mixed Hodge modules.

Let X be a complex manifold and $\mathscr{D}_X$ the sheaf of rings of holomorphic differential operators on X. Then $\mathscr{D}_X$ has the filtration F by the order of the differential operator, and $\mathrm{Gr}^F \mathscr{D}_X := \bigoplus_i \mathrm{Gr}_i^F \mathscr{D}_X$ is identified with the subsheaf of $\pi_* \mathscr{O}_{T^*X}$ consisting of sections whose restrictions to fibers $T_x^* X$ are algebraic, where $\pi : T^*X \to X$ is the cotangent bundle.

Let M be a coherent $\mathscr{D}_X$-module. We say that an exhaustive filtration F of M is *good*, if $F_i \mathscr{D}_X F_j M \subset F_{i+j} M$ and $\mathrm{Gr}^F M := \bigoplus_i \mathrm{Gr}_i^F M$ is a coherent $\mathrm{Gr}^F \mathscr{D}_X$-module. In this case, the support of $\mathrm{Gr}^F M$ is defined in T^*X and is called the characteristic variety $\mathrm{Ch}(M)$ of X. This is independent of the choice of F and depends only on M. (Note that F exists at least locally.) We say that a $\mathscr{D}_X$-module M is *holonomic* if M is coherent and $\dim \mathrm{Ch}(M) = \dim X$, and a filtered $\mathscr{D}_X$-module (M, F) is *holonomic* if M is holonomic and F is a good filtration.

For a $\mathscr{D}_X$-module M, we define the de Rham complex $\mathrm{DR}(M)$ by $\Omega_X^\cdot \otimes M[\dim X]$ whose differential is defined in the same way as the de Rham complex associated with a vector bundle with a connection. By Kashiwara [23], we have

(3.1) $\quad\quad\quad \mathrm{DR}(M) \in \mathrm{Perv}(\mathbb{C}_X) \quad \text{if} \quad M \quad \text{is holonomic,}$

where $\mathrm{Perv}(\mathbb{C}_X)$ is the category of perverse sheaves with complex coefficients, which is the full subcategory of $D_c^b(\mathbb{C}_X)$ defined by the condition :

(3.2) $\quad\quad\quad \dim \mathrm{supp}\, \mathscr{H}^i K \leq -i, \quad \dim \mathrm{supp}\, \mathscr{H}^i \mathbb{D} K \leq -i.$

See [3]. Here $\mathbb{D}$ is Verdier's dual functor [63], and $D_c^b(\mathbb{C}_X)$ is defined in the same way as $D_c^b(\mathbb{Q}_X)$ in (2.1), where the stratification is assumed analytic.

Let $f : X \to Y$ be a morphism of complex manifolds. For a $\mathscr{D}_X$-module M, we define the direct image by

$$f_* M = \mathbf{R} f_*(\mathscr{D}_{Y \leftarrow X} \otimes^{\mathbb{L}}_{\mathscr{D}_X} M)$$

(see [24]), where $\mathbf{R} f_*$ is the derived functor of the sheaf-theoretic direct image, and $\mathscr{D}_{Y \leftarrow X}$ has a structure of left $f^{-1}\mathscr{D}_Y$-module and right $\mathscr{D}_X$-module and is defined by

$$\omega_X \otimes_{f^{-1}\mathscr{O}_Y} f^{-1}(\mathscr{D}_Y \otimes_{\mathscr{O}_Y} \omega_Y^\vee).$$

Here $\omega_X = \Omega_X^{\dim X}$, and $\omega_Y^\vee$ is the dual line bundle of ω_Y. These are mainly used for the transformation of left and right $\mathscr{D}$-modules and do not appear in the definition of direct image for right $\mathscr{D}$-modules (where ω_X is replaced by $\mathscr{O}_X$).

For a filtered $\mathscr{D}$-module, the direct image is defined as follows (see also [8, 33]). It is enough to consider the case of closed embedding or smooth projection, by factorizing f using the graph of f.

For a closed embedding $i : X \to Y$, we take a local coordinate system $(y_1, \ldots, y_m)$ of Y such that $X = \{y_i = 0 \ (i \leq d)\}$, where $d = \operatorname{codim} X$. Let $\partial_i = \partial/\partial y_i$. Then ω_X, ω_Y are trivialized using the coordinates, and we have an isomorphism

$$i_* M = M \otimes_{\mathbb{C}} \mathbb{C}[\partial_1, \ldots, \partial_d].$$

If M has a filtration F, the filtration F on $i_* M$ is defined by

(3.3) $$F_p(i_* M) = \sum_v F_{p-|v|-d} M \otimes \partial^v,$$

where $v = (v_1, \ldots, v_d), |v| = \sum v_i$, and $\partial^v = \prod_i \partial_i^{v_i}$.

For a smooth projection $p : X \times Y \to Y$, the direct image $p_* M$ is expressed by

$$p_* M = \mathbf{R} p_*(\Omega^{\cdot}_{X \times Y/Y} \otimes_{\mathscr{O}} M)[\dim X],$$

using the relative differential forms $\Omega^{\cdot}_{X \times Y/Y}$. If M has a filtration F, the filtration on $p_* M$ is defined by the formula as in (1.5). Here the sheaf-theoretic direct image $\mathbf{R} p_*$ is defined by taking a functorially defined flasque resolution (whose component induces an exact functor).

So the direct image $f_*(M, F)$ is defined in general. See [39] for the definition of direct image that does not use the factorization of f.

If f is proper and M is a holonomic $\mathscr{D}_X$-module with a good filtration, the cohomological direct image $\mathscr{H}^i f_* M$ is holonomic [24]. We have a similar assertion for a filtered $\mathscr{D}_X$-module (see for example [39]), and we can define the cohomological direct image $\mathscr{H}^i f_*(M, F)$ even if the filtration of $f_*(M, F)$ is not strict. Here we say that a filtration on a complex (K, d) is *strict*, if $\mathrm{d}(F_p K^i) = F_p K^{i+1} \cap \mathrm{d} K^i$. In this case, cohomology commutes with the graduation of F. If f is proper and M is a holonomic $\mathscr{D}_X$-module, we have an isomorphism in $D^b_c(\mathbb{C}_Y)$:

(3.4) $$\mathrm{DR}(f_* M) = f_* \mathrm{DR}(M).$$

Let M be a holonomic $\mathscr{D}_X$-module. We define the dual $\mathbb{D} M$ of M by

$$\mathbf{R}\mathscr{H}om_{\mathscr{D}_X}(M, \mathscr{D}_X \otimes_{\mathscr{O}} \omega_X^\vee[\dim X]),$$

where $\mathscr{D}_X \otimes_{\mathscr{O}} \omega_X^\vee$ has a structure of left $\mathscr{D}_X$-bimodule, and $\mathbb{D} M$ is a left $\mathscr{D}_X$-module (i.e., $\mathscr{H}^i(\mathbb{D} M) = 0$ for $i \neq 0$). See [23]. We have $\mathbb{D}(\mathbb{D} M) = M$ and

(3.5) $$\mathrm{DR}(\mathbb{D} M) = \mathbb{D}(\mathrm{DR}(M)).$$

See [29, 36]. Let (M, F) be a holonomic filtered $\mathscr{D}_X$-module. We define the dual $\mathbb{D}(M, F)$ as above, where $\mathscr{D}_X \otimes_\mathscr{O} \omega_X^\vee$ has the filtration F such that

$$F_p(\mathscr{D}_X \otimes_\mathscr{O} \omega_X^\vee) = F_{p-2\dim X}\mathscr{D}_X \otimes_\mathscr{O} \omega_X^\vee.$$

Here $\mathbf{R}\mathscr{H}om$ is defined locally, for example, by taking a free resolution of (M, F). See also [39, §2]. Note that the filtration of the complex defining $\mathbb{D}(M, F)$ is not necessarily strict. In fact, it is strict if and only if (M, F) is Cohen-Macaulay (i.e., $\mathrm{Gr}^F M$ is Cohen-Macaulay over $\mathrm{Gr}^F \mathscr{D}_X$). We have $\mathbb{D}(\mathbb{D}(M, F)) = (M, F)$ if (M, F) is holonomic and Cohen-Macaulay.

Let g be a holomorphic function on X, and $i_g : X \to Y := X \times \mathbb{C}$ the embedding by the graph of g. Let $V^0\mathscr{D}_Y$ be the subring of $\mathscr{D}_Y$ generated by $\mathscr{D}_X$, $\mathscr{O}_Y$, and $t\partial_t$, where t is the coordinate of $\mathbb{C}$. For a holonomic $\mathscr{D}_X$-module M, let $\tilde{M} = (i_g)_*M$. Then, for a subset G of $\mathbb{C}$ such that $0 \in G$ and $G \to \mathbb{C}/\mathbb{Z}$ is bijective, there exists uniquely an exhaustive filtration V of $\tilde{M}$ satisfying the following conditions:
  (i) $V^i\tilde{M}$ are coherent $V^0\mathscr{D}_Y$-submodules of $\tilde{M}$.
  (ii) $tV^i\tilde{M} \subset V^{i+1}\tilde{M}$ ($i \in \mathbb{Z}$) with the equality for $i \geq 0$, $\partial_t V^i\tilde{M} \subset V^{i-1}\tilde{M}$ ($i \in \mathbb{Z}$).
  (iii) The action of $t\partial_t$ on $\mathrm{Gr}_V^i\tilde{M}$ has a (nontrivial) minimal polynomial whose roots are contained in $G + i$ locally on X.

This was obtained by Malgrange [34] in the case $M = \mathscr{O}_X$, and V is called the filtration of Kashiwara and Malgrange. The existence of V follows from [25]. We say that M is *quasi-unipotent along $g = 0$* if the roots of the minimal polynomial in condition (iii) are rational numbers. In this case, V may be indexed by $\mathbb{Q}$, where condition (iii) is replaced by
  (iii)' The action of $t\partial_t - \alpha$ on $\mathrm{Gr}_V^\alpha \tilde{M}$ is nilpotent locally on X.

Let (M, F) be a holonomic filtered $\mathscr{D}_X$-module and let $(\tilde{M}, F) = (i_g)_*(M, F)$. If M is quasi-unipotent along $g = 0$, we take the filtration V indexed by $\mathbb{Q}$, and define

$$(3.6) \quad \psi_g(M, F) = \bigoplus_{-1 < \alpha \leq 0} \mathrm{Gr}_V^\alpha(\tilde{M}, F), \quad \varphi_{g,1}(M, F) = \mathrm{Gr}_V^{-1}(\tilde{M}, F[-1]),$$

where $(F[m])_i = F_{i-m}$. In the nonfiltered case, we can define

$$\psi_g M = \mathrm{Gr}_V^0 \tilde{M}, \quad \varphi_g M = \mathrm{Gr}_V^{-1} \tilde{M},$$

where V is indexed by $\mathbb{Z}$ choosing G. But this definition does not work in the filtered case.

If M is regular holonomic [29, 35], we have

$$(3.7) \quad \mathrm{DR}(\psi_g M) = {}^p\psi_g \mathrm{DR}(M), \quad \mathrm{DR}(\varphi_{g,1} M) = {}^p\varphi_{g,1}\mathrm{DR}(M)$$

by Kashiwara [28] and Malgrange [34], where ${}^p\psi_g = \psi_g[-1]$, ${}^p\varphi_{g,1} = \varphi_{g,1}[-1]$, and $\varphi_{g,1}$ is the direct factor of φ_g with unipotent monodromy. (In this paper, Deligne's nearby and vanishing cycle functors [14] are denoted by ψ_g, φ_g.) The proof of (3.7) is not so difficult if we combine the theory of regular holonomic $\mathscr{D}$-modules with Nilson class functions.

To make the definition (3.6) useful, it is desirable that the information of $\mathrm{Gr}_V^\alpha(\tilde{M}, F)$ is not lost for $\alpha \notin [-1, 0]$. We say that (M, F) is *regular along* $g = 0$, if the following conditions are satisfied:

(3.8) $tF_p V^\alpha \tilde{M} = F_p V^{\alpha+1} \tilde{M}$ $(\alpha > -1)$, $\partial_t F_p \mathrm{Gr}_V^\alpha \tilde{M} = F_{p+1} \mathrm{Gr}_V^{\alpha-1} \tilde{M}$ $(\alpha < 0)$.

(3.9) $\mathrm{Gr}^F \mathrm{Gr}_V^\alpha \tilde{M}$ are coherent over $\mathrm{Gr}^F \mathscr{D}_X$.

If (M, F) is quasi-unipotent and regular along $g = 0$, we can construct the isomorphism (3.7) (see [39]). If (M, F) satisfies (3.8) and

$$\partial_t F_p \mathrm{Gr}_V^0 \tilde{M} = F_{p+1} \mathrm{Gr}_V^{-1} \tilde{M},$$

then we have

$$F_p \tilde{M} = \sum_{i \geq 0} \partial_t^i (V^{>-1} \tilde{M} \cap j_* j^* F_{p-i} \tilde{M})$$

(see [39, 3.2.2]), where $j : X \times \mathbb{C}^* \to X \times \mathbb{C}$ denotes a natural inclusion.

§4. Construction of Hodge modules

In this section, we explain the idea of the proof of Theorem 2.1.

In the construction of Hodge modules and the proof of their stability by direct image, the key point is the use of the functors ψ, φ. As the simplest example, consider the case of Steenbrink [57]. His argument can be extended to the filtered Gauss-Manin system $f_*(\mathscr{O}_X, F)$, and we get the proof of the decomposition theorem by showing the commutativity of the direct image with the functors ψ, φ. See [37]. This argument was inspired by Varchenko's theory of asymptotic mixed Hodge structure [61], and we used the filtration V on the Gauss-Manin system, which corresponds to the asymptotic expansion of period integrals.

To extend this argument to the case $\dim S > 1$, it is enough to take the functors ψ, φ induced by the filtration V of Kashiwara and Malgrange along the pull-back of hypersurfaces on S inductively, so that the dimension of the support decreases, and we can prove inductively the commutativity of ψ, φ with the direct image by generalizing the above argument. But we have some technical difficulties to realize this idea: e.g., the pull-back of a hypersurface must be a divisor with normal crossings (see [41]).

So we change the strategy, and try to find a category to which applies the above argument on the commutativity of ψ, φ with direct image. By analogy with the case of Steenbrink, this category should be stable by the functors $\mathrm{Gr}_i^W \psi, \mathrm{Gr}_i^W \varphi$. Let us define the category by the stability by these functors. If we can show that the functors ψ, φ commute with the direct image and the monodromy weight filtration is stable by the direct image, then the stability by the functors $\mathrm{Gr}_i^W \psi, \mathrm{Gr}_i^W \varphi$ is stable by the direct image, and we get the stability of the category by the direct image. This is the principal idea of the Hodge modules. A more precise statement is as follows.

Let X be a smooth algebraic variety and let $\mathscr{D}_X$ be the sheaf of rings of holomorphic differential operators on X. We denote by $\mathrm{MF}_h(\mathscr{D}_X, \mathbb{Q})$ the category consisting of (M, F, K) where (M, F) is a holonomic filtered $\mathscr{D}_X$-module and $K \in \mathrm{Perv}(\mathbb{Q}_X)$ with a given isomorphism $\alpha : \mathrm{DR}(M) \simeq K \otimes_{\mathbb{Q}} \mathbb{C}$. The morphisms are pairs of morphisms of (M, F) and K commuting with α. Here we assume that the stratification of K is algebraic. We say that (M, F) is given a rational structure by K and the isomorphism α. We will denote by $\mathrm{MFW}_h(\mathscr{D}_X, \mathbb{Q})$ the category consisting of objects of $\mathrm{MF}_h(\mathscr{D}_X, \mathbb{Q})$ with a finite increasing filtration W (i.e., W is a pair of filtrations of M and K compatible with α).

We define $\mathrm{MH}(X, n)$, the category of Hodge modules of weight n, by the largest full subcategory of $\mathrm{MF}_h(\mathscr{D}_X, \mathbb{Q})$ satisfying the following three conditions:
  (i) Any object has the decomposition by strict support (2.4).
  (ii) For any function g defined on an open subvariety U of X, any object (M, F, K) is quasi-unipotent and regular along $g = 0$ and $\mathrm{Gr}_i^W \psi_g(M, F, K)|_U$, $\mathrm{Gr}_i^W \varphi_{g,1}(M, F, K)|_U \in \mathrm{MH}(U, i)$, where W is the monodromy filtration shifted by $n-1$ and n respectively.
  (iii) Any object whose support is a point x is isomorphic to $(i_x)_*(H_\mathbb{C}, F, H)$ for (H, F) a Hodge structure of weight n, where $i_x : \{x\} \to X$ is a natural inclusion and we put $F_p = F^{-p}$.

Here $\psi_g, \varphi_{g,1}$, and $(i_x)_*$ are defined using (3.7) and (3.4). By (i), we may assume that (M, F, K) in (ii) has strict support Z. If $g|_Z = 0$, we have $\psi_g(M, F, K)|_U = 0$, $\varphi_{g,1}(M, F, K)|_U = (M, F, K)|_U$, the monodromy filtration is trivial (i.e., $\mathrm{Gr}_i^W = 0$ for $i \neq n$), and the condition of regularity is equivalent to: $g(F_p M|_U) \subset F_{p-1} M|_U$ (see [39, 3.2.6]). If $g|_Z \neq 0$, the dimension of the support decreases, and we can define $\mathrm{MH}(X, n)$ by induction on dimension. Note that the above condition is Zariski-local on X.

We can show that the underlying $\mathscr{D}_X$-module of a Hodge module is regular holonomic [29, 35] by using the condition (ii), and we may assume that the $\mathscr{D}_X$-modules are regular holonomic from the beginning. The regularity of $\mathscr{D}$-modules is used in an essential way to show the uniqueness of the underlying $\mathscr{D}$-module of the direct image of a mixed Hodge module by an open embedding.

We now define a *polarization* of Hodge modules. Let (M, F, K) be a Hodge module of weight n. A polarization of (M, F, K) is a pairing $S : K \otimes K \to \mathbb{D}_X(-n)$ (where $\mathbb{D}_X$ is the dualizing sheaf [63]) such that the corresponding morphism $K \to (\mathbb{D}K)(-n)$ is an isomorphism and is lifted to an isomorphism $(M, F, K) \xrightarrow{\sim} (\mathbb{D}(M, F, K))(-n)$, and the following two conditions are satisfied:
  (i) For any function g in the above condition (ii) for Hodge modules, $\psi_g S \circ (\mathrm{id} \otimes N^i)$ gives a polarization on the primitive part $P\mathrm{Gr}_{n-1+i}^W \psi_g(M, F, K)|_U$ $(:= \mathrm{Ker}\, N^{i+1})$.
  (ii) If the support is a point $\{x\}$, S corresponds to a polarization of Hodge structure [12] by the isomorphism in the above condition (iii).

Here the dual $\mathbb{D}(M, D, K)$ is defined using (3.5), and the Tate twist (r) for an integer r is defined by

$$(M, F)(r) = (M, F[r]), \quad K(r) = K \otimes \mathbb{Z}(r),$$

where $(F[r])_p = F_{p-r}$, and $\mathbb{Z}(r) = (2\pi i)^r \mathbb{Z} \subset \mathbb{C}$. Note that the Tate twist for a $\mathbb{C}$-module is trivialized using the natural isomorphism $\mathbb{C} \otimes \mathbb{Z}(r) \xrightarrow{\sim} \mathbb{C}$ (forgetting the filtration). See [39, 5.2] for the definition of $\psi_g S$. The condition for polarization is also given by induction on dimension. We will denote by $\mathrm{MH}(X, n)^p$ the category of polarizable Hodge modules of weight n.

THEOREM 4.1. *Let $f : X \to Y$ be a projective morphism of complex manifolds, and l the first Chern class of a relatively ample line bundle. Then, for $(M, F, K) \in \mathrm{MH}(X, n)^p$,* the direct image $f_*(M,F)$ is strict, $\mathcal{H}^i f_*(M,F,K) := (\mathcal{H}^i f_*(M,F), {}^p\mathcal{H}^i f_*K)$ belongs to $\mathrm{MH}(Y, n+i)^p$, and the hard Lefschetz property

(4.1) $\qquad l^i : \mathcal{H}^{-i} f_*(M,F,K) \xrightarrow{\sim} \mathcal{H}^i f_*(M,F,K)(i)$

holds. Furthermore, for a polarization S of (M,F,K), $(-1)^{i(i-1)/2} f_* S \circ (\mathrm{id} \otimes l^i)$ gives a polarization on the primitive part $P\mathcal{H}^{-i} f_*(M,F,K) := \mathrm{Ker}\, l^{i+1}$.

Here ${}^p\mathcal{H}^i$ is the perverse cohomology functor [3] such that $\mathrm{DR} \circ \mathcal{H}^i = {}^p\mathcal{H}^i \circ \mathrm{DR}$, and $\mathcal{H}^i f_*(M,F,K)$ is defined using (3.4). Theorem 4.1 is proved by induction on dimension, taking the graduation of the filtration V associated with (the pull-back of) a locally defined function on Y. In the case $Y = \mathrm{pt}$, we take a Lefschetz pencil so that the assertion is reduced to the case $X = \mathbb{P}^1$ and follows from the result of Zucker [65]. See [39]. The decomposition by strict support in the condition (i) for Hodge modules is equivalent to

(4.2) $\qquad\qquad\qquad \varphi_{g,1} = \mathrm{Im\, can} \oplus \mathrm{Ker\, Var}$

for functions g in condition (ii) and is proved using a polarization. See [39] for can, Var.

We define the category of mixed Hodge modules $\mathrm{MHM}(X)$ by the largest full subcategory of $\mathrm{MFW}_h(\mathcal{D}_X, \mathbb{Q})$ such that $\mathrm{Gr}_i^W \mathcal{M} \in \mathrm{MH}(X,i)^p$ for $\mathcal{M} \in \mathrm{MHM}(X)$ and $\mathrm{MHM}(X)$ is stable by the functors $j_!$, j_*, ψ_g, $\varphi_{g,1}$, $\boxtimes \mathbb{Q}_Y^H[\dim Y]$ and also by the pull-back to open subvarieties, where j is the inclusion of the complement of a divisor, g is a function, Y is a smooth variety, and $\mathbb{Q}_Y^H[\dim Y]$ is defined by

$(\mathcal{O}_Y, F, \mathbb{Q}_Y[\dim Y]; W) \quad \text{with} \quad \mathrm{Gr}_p^F = 0 \ (p \neq 0), \mathrm{Gr}_i^W = 0 \ (i \neq \dim Y).$

Here we assume the existence of a relative monodromy filtration [15] and the compatibility [39, §1] of the three filtrations F, W, V when we consider the functors $\psi_g, \varphi_{g,1}$. These conditions are a natural generalization of the condition for admissible variation of mixed Hodge structure by Steenbrink-Zucker [60]. We can also define the open direct images $j_!, j_*$ extending the arguments in [loc. cit.], but we have to choose (locally) a function g which vanishes exactly on the complement of $\mathrm{Im}\, j$, and we have to assume the independence of $j_!, j_*$ by the choice of g. See [40, 4.2] for the details.

So $\mathrm{MHM}(X)$ is defined in the X smooth case. In general, it is defined by using a locally-defined closed embedding of X into smooth varieties. See [40, 53]. For a projective morphism f, the stability of mixed Hodge modules by the cohomological direct image $\mathcal{H}^i f_*$ follows from Theorem 4.1 together with the stability of the monodromy weight filtration by the direct image. Using this and the stability by external product, it is not difficult to get the other functors in Theorem 2.1. See [40].

§5. Application

In this section, we explain some of the applications of the theory of mixed Hodge modules.

(i) Let (L, F) be a polarizable variation of Hodge structure defined on a smooth Zariski-open subset U of an irreducible variety X. Let $p = \max\{p : F^p \mathscr{L} \neq 0\}$, where $\mathscr{L} = L \otimes_\mathbb{Q} \mathcal{O}_U$ (see §1). Then Kollár's conjecture states that $\omega_U \otimes F^p \mathscr{L}$ is naturally extended to a coherent sheaf on X, and this procedure commutes with the direct image by a proper morphism. This can be checked using Theorems 2.8 and 4.1, etc. See [45] for the details. Here it is natural to use right $\mathcal{D}$-modules.

As to the vanishing theorem, we have the following (see [40]):

THEOREM 5.1. *Let X be a projective variety with an ample line bundle L, and $\mathcal{M} \in \mathrm{MHM}(X)$. Assume X is embedded in $Y = \mathbb{P}^N$ by a power of L so that $\mathcal{M}$ is represented by $(M, F, K; W)$ on Y. Then $\mathrm{Gr}_p^F \mathrm{DR}_Y(M)$ is a complex of $\mathcal{O}_X$-modules and*

$$H^i(X, \mathrm{Gr}_p^F \mathrm{DR}_Y(M) \otimes L) = 0 \ (i > 0), \quad H^i(X, \mathrm{Gr}_p^F \mathrm{DR}_Y(M) \otimes L^{-1}) = 0 \ (i < 0).$$

(ii) Let Z be a closed subvariety of an algebraic variety X. The *link* L of Z in X is defined by $g^{-1}(\varepsilon)$ for $0 < \varepsilon \ll 1$, where g is an appropriate nonnegative-valued continuous function that vanishes exactly on Z. The intersection cohomology $\mathrm{IH}^{\cdot}(L)$ (and also $\mathrm{IH}_c^i(L)$ with compact supports) has a natural mixed Hodge structure, and $\mathrm{IH}^i(L)$ (resp. $\mathrm{IH}_c^i(L)$) has weights $> i$ (resp. weights $\leq i$) for $i \geq \dim X + \dim Z$ (resp. $i < \dim X - \dim Z$). This implies, for example, the vanishing of the cup product in some cases, and a real torus cannot be the link of Z for $\dim X - \dim Z > 1$. See [19].

(iii) Let X be a smooth algebraic variety, Y a reduced divisor on X, and $U = X \setminus Y$ with a natural inclusion $j : U \to X$. Let $\mathcal{O}_X(*Y) = \bigcup_n \mathcal{O}_X(nY)$ (i.e., the sheaf of meromorphic functions whose poles are contained in Y). Then $\mathcal{O}_X(*Y)$ is the underlying $\mathcal{D}_X$-module of the mixed Hodge module $j_*\mathbb{Q}_U^H[\dim U]$ and has the Hodge filtration F. We define the pole order filtration P by

$$P_i \mathcal{O}_X(*Y) = \mathcal{O}_X((i+1)Y) \quad \text{for} \quad i \geq 0 \quad \text{and} \quad 0 \quad \text{otherwise}.$$

See [18]. Then we have

PROPOSITION 5.2. $F_i \mathcal{O}_X(*Y) \subset P_i \mathcal{O}_X(*Y)$.

In fact, they coincide on $X \setminus \mathrm{Sing}\, Y$, and we get the assertion, because $F_i \mathcal{O}_X(*Y)$ is torsion-free and $P_i \mathcal{O}_X(*Y)$ is locally free. Note that 5.2 implies a result of [18]. Let $b_f(s)$ be the b-function of a local equation f of D, and $-\alpha_f$ the maximal root of $b_f(s)/(s+1)$. Then we have the equality $F_i \mathcal{O}_X(*Y) = P_i \mathcal{O}_X(*Y)$ for $i \leq \alpha_f - 1$. (This is inspired by Deligne's result.)

(iv) It has been expected that the theory of mixed Hodge modules is related to that of mixed motives. Let $\mathrm{MHM}(X)^{\mathrm{go}}$ denote the full subcategory consisting of mixed Hodge modules of geometric origin (cf. [3]). For a smooth projective variety X, we have a cycle class map

(5.1) $$\mathrm{cl}^{\mathrm{MH}} : \mathrm{CH}^p(X)_{\mathbb{Q}} \to \mathrm{Ext}^{2p}(\mathbb{Q}_X^H, \mathbb{Q}_X^H(p)),$$

where $\mathrm{CH}^p(X)_{\mathbb{Q}}$ is the Chow group with rational coefficients, and Ext is taken in the bounded derived category of $\mathrm{MHM}(X)^{\mathrm{go}}$. If the Hodge conjecture is true for any smooth projective variety, (5.1) is surjective. The injectivity is related to the surjectivity of the cycle map of Bloch's higher Chow group $\mathrm{CH}^p(X, 1)_{\mathbb{Q}}$ [5]. See [50]. To avoid the Hodge conjecture, we can introduce the notion of geometric level. In fact, (5.1) is bijective for $p \leq 2$ if Ext is taken in the bounded derived category of the full subcategory consisting of mixed Hodge modules of geometric level $\leq \dim X$ (see [loc. cit.]).

For an algebraic variety X over an algebraic number field k, we have to consider the combination with an l-adic perverse sheaf on $X \otimes_k \overline{k}$ with the action of $\mathrm{Gal}(\overline{k}/k)$. This is inevitable if we want to lower the field of definition of a cycle.

For other applications such as the conjectures of Illusie and Steenbrink, see [55, 49], etc.

References

1. A. Beilinson, *Height pairing between algebraic cycles*, Lecture Notes in Math., vol. 1289, Springer-Verlag, Berlin, 1987, pp. 1–26.
2. _____, *On the derived category of perverse sheaves*, Lecture Notes in Math., vol. 1289, Springer-Verlag, Berlin, 1987, pp. 27–41.
3. A. Beilinson, J. Bernstein, and P. Deligne, *Faisceaux pervers*, Astérisque **100**, Soc. Math. France, Paris, 1982.
4. S. Bloch, *Lectures on algebraic cycles*, Duke University Mathematical Series 4, Durham, 1980.
5. _____, *Algebraic cycles and higher K-theory*, Advances in Math. **61** (1986), 267–304.
6. S. Bloch and A. Ogus, *Gersten's conjecture and the homology of schemes*, Ann. Sci. École Norm. Sup. (4) **7** (1974), 181–201.
7. A. Borel, *Algebraic $\mathcal{D}$-modules*, Academic Press, Boston, 1987.
8. J.-L. Brylinski, *Modules holonomes à singularités régulières et filtrations de Hodge*. II, Astérisque **101–102** (1983), 75–117.
9. E. Cattani and A. Kaplan, *Polarized mixed Hodge structures and the local monodromy of a variation of Hodge structure*, Invent. Math. **67** (1982), 101–115.
10. E. Cattani, A. Kaplan, and W. Schmid, L^2 *and intersection cohomologies for a polarizable variation of Hodge structure*, Invent. Math. **87** (1987), 217–252.
11. J. Cheeger, M. Goresky, and R. MacPherson, L^2-*cohomology and intersection homology of singular algebraic varieties*, Seminar on Differential Geometry, Ann. of Math. Studies, vol. 102, Princeton Univ. Press, Princeton, NJ, 1982, pp. 303–340.
12. P. Deligne, *Théorie de Hodge*. I, Inst. Hautes Études Sci. Actes Congrès. Intern. Math., (1970), 425–430; II, Publ. Math. **40** (1971), 5–58; III, ibid. **44** (1974), 5–77.
13. _____, *Équation différentielle à points singuliers réguliers*, Lecture Notes in Math., vol. 163, Springer-Verlag, Berlin, 1970.
14. _____, *Le formalisme des cycles évanescents*, in SGA7 XIII and XIV, Lecture Notes in Math., vol. 340, Springer-Verlag, Berlin, 1973, pp. 82–115 and 116–164.
15. _____, *Conjecture de Weil*. II, Inst. Hautes Études Sci. Publ. Math. **52** (1980), 137–252.
16. _____, *Valeurs de fonctions L et périodes d'intégrales*, in Proc. Sympos. Pure Math., vol. 33, part 2, Amer. Math. Soc., Providence, RI, 1979, pp. 313–346.
17. _____, *Le groupe fondamental de la droite projective moins trois points*, Galois Groups over **Q**, Springer-Verlag, New York, 1989, pp. 79–297.
18. P. Deligne and A. Dimca, *Filtrations de Hodge et par l'ordre du pôle pour les hypersurfaces singulières*, Ann. Sci. École Norm. Sup. (4) **23** (1990), 645–656.
19. A. Durfee and M. Saito, *Mixed Hodge structure on the intersection cohomology of links*, Compositio Math. **76** (1990), 49–67.
20. P. Griffiths, *Periods of integrals on algebraic manifolds*. I, Amer. J. Math. **90** (1968), 568–626; II, ibid., 805–865; III, Inst. Hautes Études Sci. Publ. Math. **38** (1970), 125–180.
21. U. Jannsen, *Deligne homology, Hodge D-conjecture, and motives*, Beilinson's Conjecture on Special Values of L-functions, Academic Press, Boston, 1988, pp. 305–372.
22. _____, *Mixed motives and algebraic K-theory*, Lecture Notes in Math., vol. 1400, Springer-Verlag, Berlin, 1990.
23. M. Kashiwara, *On the maximally overdetermined system of linear differential equations*. I, Publ. Res. Inst. Math. Sci., Kyoto Univ., 1975, pp. 563–579.
24. _____, *B-function and holonomic systems*, Invent. Math. **38** (1976), 33–53.
25. _____, *On the holonomic systems of differential equations*. II, Invent. Math. **49** (1978), 121–135.
26. _____, *Quasi-unipotent constructible sheaves*, J. Fac. Sci. Univ. Tokyo **28** (1981), 757–773.
27. _____, *A study of variation of mixed Hodge structure,*, Publ. Res. Inst. Math. Sci., vol. 22, Kyoto Univ., 1986, pp. 991–1024.
28. _____, *Vanishing cycle sheaves and holonomic systems of differential equations*, Lecture Notes in Math., vol. 1016, Springer-Verlag, Berlin, 1983, pp. 136–142.
29. M. Kashiwara and T. Kawai, *On the holonomic system of microdifferential equations*. III, Publ. Res. Inst. Math. Sci., vol. 17, Kyoto Univ., 1981, pp. 813–979.
30. _____, *The Poincaré lemma for variations of polarized Hodge structures*, Publ. Res. Inst. Math. Sci., vol. 23, Kyoto Univ., 1987, pp. 345–407.

31. _____, *Hodge structure and holonomic systems*, Proc. Japan Acad. Ser. A Math. Sci. **62** (1986), 1–4.
32. J. Kollár, *Higher direct images of dualizing sheaves.* I, Ann. of Math. (2) **123** (1986), 11–42; II, ibid. **124** (1986), 171–202.
33. G. Laumon, *Sur la catégorie dérivée des $\mathscr{D}$-modules filtrés*, Lecture Notes in Math., vol. 1016, Springer-Verlag, Berlin, 1983, pp. 151–237.
34. B. Malgrange, *Polynôme de Bernstein-Sato et cohomologie évanescente*, Astérisque **101–102** (1983), 243–267.
35. Z. Mebkhout, *Une autre équivalence de catégories*, Compositio Math. **51** (1984), 63–88.
36. _____, *Théorème de bidualité pour les $\mathscr{D}_X$-modules holonomes*, Arkiv Mat. **20** (1982), 111–124.
37. M. Saito, *Hodge filtration on Gauss-Manin systems.* I, J. Fac. Sci. Univ. Tokyo, Sect. I A Math. **30** (1984), 489–498; II, Proc. Japan Acad. Ser. A Math. Sci. **59** (1983), 37–40.
38. _____, *Hodge structure via filtered $\mathscr{D}$-modules*, Astérisque **130** (1985), 342–351.
39. _____, *Modules de Hodge polarisables*, Publ. Res. Inst. Math. Sci., vol. 24, Kyoto Univ., 1988, pp. 849–995.
40. _____, *Mixed Hodge modules*, Publ. Res. Inst. Math. Sci., vol. 26, Kyoto Univ., 1990, pp. 221–333.
41. _____, *Decomposition theorem for proper Kähler morphisms*, Tôhoku Math. J. (2) **42** (1990), 127–148.
42. _____, *Duality for vanishing cycle functors*, Publ. Res. Inst. Math. Sci., vol. 25, Kyoto Univ., 1989, pp. 889–921.
43. _____, *Extension of mixed Hodge modules*, Compositio Math. **74** (1990), 209–234.
44. _____, *Hodge conjecture and mixed motives.* I, Proc. Sympos. Pure Math., vol. 53, Amer. Math. Soc., Providence, RI, 1991, pp. 283–303; II, in Lecture Notes in Math., Springer-Verlag, Berlin, vol. 1479, 1991, pp. 196–215.
45. _____, *On Kollár's conjecture*, Proc. Sympos. Pure Math., vol. 52, Amer. Math. Soc., Providence, RI, 1991, Part 2, pp. 509–517.
46. _____, *Vanishing cycles and mixed Hodge modules*, preprint, Inst. Hautes Études Sci., 1988.
47. _____, *Introduction to mixed Hodge modules*, Astérisque **179–180** (1989), 145–162.
48. _____, *Mixed Hodge modules and admissible variations*, C. R. Acad. Sci. Paris Sér. I Math. **309** (1989), 351–356.
49. _____, *On Steenbrink's conjecture*, Math. Ann. **289** (1991), 703–716.
50. _____, *On the injectivity of cycle maps*, Publ. Res. Inst. Math. Sci., vol. 28, Kyoto Univ., 1992, pp. 99–127.
51. _____, *On the formalism of mixed sheaves*, Res. Inst. Math. Sci., vol. 784, Kyoto Univ., 1991.
52. _____, *Induced $\mathscr{D}$-modules and differential complexes*, Bull. Soc. Math. France **117** (1989), 361–387.
53. _____, *$\mathscr{D}$-modules on analytic spaces*, Publ. Res. Inst. Math. Sci., vol. 27, Kyoto Univ., 1991, pp. 291–332.
54. _____, *Mixed Hodge Modules and applications*, Proc. Internat. Congr. Math. Kyoto 1990, Springer, Tokyo, 1991, pp. 725–734.
55. M. Saito and S. Zucker, *The kernel spectral sequence of vanishing cycles*, Duke Math. J. **61** (1990), 329–339.
56. W. Schmid, *Variation of Hodge structure*: *The singularity of period mapping*, Invent. Math. **22** (1973), 211–319.
57. J. Steenbrink, *Limits of Hodge structures*, Invent. Math. **31** (1976), 229–257.
58. _____, *Mixed Hodge structure on the vanishing cohomology*, Real and Complex Singularities, Oslo 1976, Sijthoff-Noordhoff, Alphen an der Rijn, 1977, pp. 525–563.
59. _____, *The spectrum of hypersurface singularity*, Astérisque **179–180** (1989), 163–184.
60. J. Steenbrink and S. Zucker, *Variation of mixed Hodge structure.* I, Invent. Math. **80** (1985), 489–542.
61. A. Varchenko, *Asymptotic Hodge structure in the vanishing cohomology*, Math. USSR-Izv. **18** (1982), 469–512.
62. J.-L. Verdier, *Catégories dérivées, Etat 0*, in SGA 4 1/2, Lecture Notes in Math., vol. 569, Springer-Verlag, Berlin, 1977, pp. 262–308.
63. _____, *Dualité dans les espaces localement compacts*, Séminaire Bourbaki, no. 300 (1965/66).
64. _____, *Extension of a perverse sheaf over a closed subspace*, Astérisque **130** (1985), 210–217.
65. S. Zucker, *Hodge theory with degenerating coefficients*, L_2-*cohomology in the Poincaré metric*, Ann. of Math. (2) **109** (1979), 415–476.

RESEARCH INSTITUTE FOR MATHEMATICAL SCIENCES, KYOTO UNIVERSITY, KYOTO 606, JAPAN

Translated by M. SAITO

Integral Invariants in Kähler Geometry

Akito Futaki

§1. Introduction

One of the fundamental problems in differential geometry is to give differentiable manifolds metrics that have "good" properties. In Kähler geometry one way to find Kähler metrics with certain good properties is to search for the Kähler forms within a given de Rham cohomology class.

"Integral invariants" in the title are invariants that are defined in terms of certain integrals by using Kähler forms but the invariants depend only on the Kähler class (namely, the cohomology classes represented by the Kähler forms). We shall see in this article that such integral invariants are defined by using holomorphic vector fields and that the invariants become obstructions to the existence of certain good metrics. In particular, we shall see that one of these invariants becomes an obstruction to the existence of Kähler-Einstein metrics in the case where M is a compact complex manifold with $c_1(M) > 0$ and the Kähler class is equal to $c_1(M)$. Moreover, we shall see that the integral invariants described above are formulated as characters of the complex Lie algebra consisting of holomorphic vector fields on M. In §§1 to 4, these vector fields will play a pivotal role in discussing the existence and the uniqueness of good Kähler metrics.

On the other hand, we can introduce characters from classical invariants, which depend only on the complex structure of M and are expressed by the integrals using Hermitian metrics. The integral invariants in the case where $c_1(M) > 0$, which was stated above, may be regarded as an example of such characters. In other words, there are two families of invariants: namely, the family of invariants that depend only on the Kähler class and the family of invariants that depend only on the complex structure. When $c_1(M) > 0$, the invariants stated above are considered as belonging to the former when $c_1(M)$ is regarded as a Kähler class and to the latter when the property that $c_1(M)$ depends only on the complex structure is emphasized. We shall regard the invariants as belonging to the latter from §5 onward. As for reference material, see the references in [20, 18] in addition to the references at the end of this paper.

1991 *Mathematics Subject Classification.* Primary 53C55.

This article originally appeared in Japanese in Sūgaku **44** (1) (1992), 44–56.

§2. Extremal Kähler metrics

Let M be a compact m-dimensional complex manifold, and let $T'M$ and $T''M$ be the holomorphic and antiholomorphic tangent bundles, respectively. A smooth cross section g of $T'M^* \otimes T''M^*$ is called a Hermitian metric if it gives a Hermitian inner product of $T'_p M$ for each $p \in M$. Then the matrix $(g_{\alpha\bar{\beta}})$ is a nonsingular Hermitian matrix where

$$g_{\alpha\bar{\beta}} = g\left(\frac{\partial}{\partial z^\alpha}, \frac{\partial}{\partial z^{\bar{\beta}}}\right)$$

and $(z^1, \ldots, z^m)$ are holomorphic local coordinates. Furthermore, g is said to be a Kähler metric if the $(1,1)$-form

$$\omega(g) = \frac{i}{2\pi} \sum_{\alpha,\beta} g_{\alpha\bar{\beta}} \, dz^\alpha \wedge dz^{\bar{\beta}}$$

is closed. Then (M, g) is called a Kähler manifold, $\omega(g)$ the Kähler form, and the de Rham cohomology class $[\omega(g)]$ the Kähler class.

PROPOSITION 2.1. *Assume that a real (k, k)-form α on a Kähler manifold is exact. Then there exists a real $(k-1, k-1)$-form β such that $\alpha = i\partial\bar{\partial}\beta$. Moreover, when $k = 1$, the smooth function β is unique up to a constant.*

Now we consider the case where two Kähler metrics g and g' give the same Kähler class. Then it follows from Proposition 2.1 that there exists a real function φ such that

$$\omega(g') = \omega(g) + \frac{i}{2\pi}\partial\bar{\partial}\varphi; \qquad g'_{\alpha\bar{\beta}} = g_{\alpha\bar{\beta}} + \varphi_{\alpha\bar{\beta}},$$

where

$$\varphi_{\alpha\bar{\beta}} = \frac{\partial^2 \varphi}{\partial z^\alpha \partial z^{\bar{\beta}}}.$$

The g' above is called the Kähler deformation of g. The reader who is familiar with Yamabe's Problem may understand that it corresponds to the conformal deformation. In fact, they are essentially the same when the dimension is equal to 1. In this article we shall deform a given Kähler metric to a good Kähler metric by using a Kähler deformation, where a good metric means a metric whose curvature has good properties. Now we shall define the curvature. Let $R(g)$ and $\sigma(g)$ denote the Ricci curvature and the scalar curvature of a Kähler metric g, respectively. Then $R(g)$ is a smooth cross section of $T'M^* \otimes T''M^*$, $\sigma(g)$ is a smooth function on M, and $R(g)$, $\sigma(g)$ are defined by the following:

$$R(g)_{\alpha\bar{\beta}} = -\frac{\partial^2}{\partial z^\alpha \partial z^{\bar{\beta}}} \log \det(g_{\gamma\bar{\delta}}),$$

$$\sigma(g) = \sum_{\alpha,\beta} g^{\alpha\bar{\beta}} R(g)_{\alpha\bar{\beta}},$$

where $(g^{\alpha\bar{\beta}})$ is the inverse of $(g_{\alpha\bar{\beta}})$ which defines a smooth cross section of $T'M \otimes T''M$. Now according to Calabi [11], we shall pose a variational problem using the terminology above. Let Ω be a fixed de Rham cohomology class of type $(1,1)$, let

Ω^+ be the set of all Kähler metrics whose Kähler classes are equal to Ω. Then a functional $\Phi\colon \Omega^+ \to \mathbb{R}^+$ is defined by

$$\Phi(g) = \int_M \sigma(g)^2 \omega(g)^m, \qquad \omega(g)^m = \omega(g) \wedge \cdots \wedge \omega(g) \quad (m \text{ times}).$$

A critical point of Φ is called an extremal Kähler metric.

PROPOSITION 2.2. *g is an extremal Kähler metric if and only if*

$$\operatorname{grad} \sigma(g) = \sum_{\alpha,\beta} g^{\alpha\bar{\beta}} \frac{\partial \sigma(g)}{\partial z^{\bar{\beta}}} \frac{\partial}{\partial z^\alpha}$$

is a holomorphic vector field.

Hence g is an extremal Kähler metric if $\sigma(g)$ is a constant function of M, and g is called a Kähler-Einstein metric (which will be abbreviated to K-E metric hereafter) if there exists a real constant c such that

$$R(g) = cg.$$

Since it follows from the condition above that

$$\sigma(g) = cm,$$

namely, that $\sigma(g)$ is a constant function, a K-E metric is necessarily an extremal Kähler metric. In this article, by good metrics we shall mean extremal Kähler metrics and, as special cases, Kähler metrics with constant scalar curvature or K-E metrics. The following theorem of Calabi gives a necessary condition for the existence of extremal Kähler metrics.

THEOREM 2.3 ([12]). *Assume that M has an extremal Kähler metric g. Then the Lie algebra $\mathfrak{H}(M)$ of all homomorphic vector fields on M has the following semiproduct structure*:
  (1) $\mathfrak{H}(M) = \mathfrak{H}_0 + \sum_{\lambda \neq 0} \mathfrak{H}_\lambda$,
  (2) $\mathfrak{H}_0$ *is reductive and* $\operatorname{grad} \sigma(g)$ *belongs to the center of* $\mathfrak{H}_0$,
  (3) $\mathfrak{H}_\lambda$ *is the λ-eigenspace of the adjoint representation of* $\operatorname{grad} \sigma(g)$.

The above theorem implies the theorem of Lichnerowicz-Matsushima ([29, 32]) which states that $\mathfrak{H}(M)$ is reductive if $\sigma(g)$ is constant. Thus, Calabi's theorem is regarded as a generalization of L-M's theorem. An example of an M whose $\mathfrak{H}(M)$ does not satisfy the conditions in Theorem 2.3 is known ([28]). On the other hand, though the conditions in Theorem 2.3 are trivially satisfied if $\mathfrak{H}(M) = 0$, there is an example M such that $\mathfrak{H}(M) = 0$ and M does not admit any extremal Kähler metric (whose scalar curvature is necessarily constant in this case) with a given Kähler class ([10]). Hence the reductivity of $\mathfrak{H}(M)$ is not a sufficient condition for the existence of Kähler metrics with constant scalar curvatures in a given Kähler class.

§3. Integral invariants

Let (M, g) be a compact Kähler manifold. It follows from the Hodge theory for the Laplacian $\Delta = \partial^2/\partial z^\alpha \partial z^{\bar\beta}$ that there exists a smooth function ϕ such that $\varphi = \Delta \phi$ if the smooth function φ satisfies $\int_M \varphi \omega(g)^m = 0$. Hence we have

$$\sigma(g) - \frac{\int_M \sigma(g)\omega(g)^m}{\int_M \omega(g)^m} = \Delta F(g) \tag{3.1}$$

for some smooth function $F(g)$. Now a linear map $f : \mathfrak{H}(M) \to \mathbb{C}$ is defined by

$$f(X) = \frac{(m+1)i}{2\pi} \int_M XF(g)\omega(g)^m, \tag{3.2}$$

where $XF(g)$ is the derivative of $F(g)$ by a vector field $X \in \mathfrak{H}(M)$. Here the Kähler class Ω is fixed, and f is defined for any $g \in \Omega^+$. Then we have the following theorem.

THEOREM 3.3 ([16, 12, 5]). *f is determined only by Ω and is independent of the choice of $g \in \Omega^+$. Moreover, f is a character of a Lie algebra (namely, a Lie algebra homomorphism), and $f = 0$ if there exists a Kähler metric of constant scalar curvature in Ω^+.*

Now we apply Theorem 3.3 to the case of K-E metrics. For a Hermitian metric g of a complex manifold, a real closed form $\rho(g)$ is defined by

$$\rho(g) = \frac{i}{2\pi} \sum_{\alpha,\beta} R(g)_{\alpha\bar\beta}\, dz^\alpha \wedge dz^{\bar\beta} = -\frac{i}{2\pi} \partial\bar\partial \log \det(g_{\alpha\bar\beta}),$$

and it is clear that the de Rham cohomology class $[\rho(g)]$ is independent of the choice of g. $[\rho(g)]$ is called the first Chern class and is denoted by $c_1(M)$, and $\rho(g)$ is called the first Chern form. If g is a K-E metric, then there exists a real constant c such that $R(g)_{\alpha\bar\beta} = c \cdot g_{\alpha\bar\beta}$. It follows from the definition of the Ricci curvature that c can be normalized to be equal to -1, 0, or 1 by changing scale. Thus there are the following three cases.

   (i) $c = -1$. In this case $\rho(g) = -\omega(g)$, and $c_1(M)$ is represented by a negative-definite closed $(1,1)$-form. We denote this case by writing $c_1(M) < 0$.
   (ii) $c = 0$. Hence, $\rho(g) = 0$ and $c_1(M) = 0$.
   (iii) $c = 1$. Hence $\rho(g) = \omega(g)$ and $c_1(M)$ is represented by a positive-definite closed $(1,1)$-form. We denote this case by writing $c_1(M) > 0$.

The problem here is whether the converse is true, namely, whether M admits a K-E metric if $c_1(M) < 0$, $c_1(M) = 0$, or $c_1(M) > 0$.

THEOREM 3.4 ([2, 40]). *If $c_1(M) < 0$, then there exists a unique K-E metric in the Kähler class $-c_1(M)$. If $c_1(M) = 0$, then there exists a unique K-E metric in each fixed Kähler class.*

Thus, we consider the case of $c_1(M) > 0$ in what follows. Since $R(g)_{\alpha\bar\beta} = g_{\alpha\bar\beta}$ and $\omega(g)$ represents $c_1(M)$ if there exists a K-E metric g with $c = 1$, we fix the Kähler class Ω so that $\Omega = c_1(M)$. Ω^+ is defined in the same way as in §2. For any $g \in \Omega^+$, it follows from Proposition 2.1 that there exists a smooth function $F(g)$ such that

$$(3.5) \qquad \rho(g) - \omega(g) = \frac{i}{2\pi} \partial\bar{\partial} F(g).$$

It is clear that the above equality is equivalent to (3.1). So $f : \mathfrak{H}(M) \to \mathbb{C}$ is defined by (3.2). Since $\Omega = c_1(M)$ depends only on the complex structure of M, it follows from Theorem 3.3 that f depends only on the complex structure of M. Furthermore, it follows from (3.5) that $f = 0$ is a necessary condition for the existence of K-E metrics on M.

EXAMPLE 3.6. Let M be the surface obtained by blowing up $\mathbb{P}^2(\mathbb{C})$ at (1:0:0). Then $c_1(M) > 0$ and $\mathfrak{H}(M)$ is isomorphic to

$$\left\{ \begin{pmatrix} * & * & * \\ 0 & * & * \\ 0 & * & * \end{pmatrix} \in \mathfrak{gl}(3; \mathbb{C}) \right\}$$

divided by the center. Since this Lie algebra is not reductive, it follows from the Lichnerowicz-Matsushima theorem that M does not admit any K-E metric. On the other hand, it can be seen from direct computations that $f(X) \neq 0$ for the vector field X corresponding to

$$\begin{pmatrix} 0 & 0 & 0 \\ 0 & 1 & 0 \\ 0 & 0 & 1 \end{pmatrix}.$$

Hence we can also deduce that M does not admit any K-E metric.

Compact complex surfaces M with $c_1(M) > 0$ are completely classified as one of $\mathbb{P}^1(\mathbb{C}) \times \mathbb{P}^1(\mathbb{C})$, $\mathbb{P}^2(\mathbb{C})$ or the blowing-up of $\mathbb{P}^2(\mathbb{C})$ at k points ($1 \leq k \leq 8$) in general position. $\mathbb{P}^1(\mathbb{C}) \times \mathbb{P}^1(\mathbb{C})$ and $\mathbb{P}^2(\mathbb{C})$ are K-E manifolds with respect to the standard metrics. As was seen in Example 3.6, there does not exist a K-E metric when $k = 1$. Neither does there exist a K-E metric when $k = 2$ since it can be seen similarly to the case $k = 1$ that $\mathfrak{H}(M)$ is not reductive and, furthermore, $f \neq 0$. When $k = 3$, we can show that $\mathfrak{H}(M)$ is reductive and $f = 0$. When $k \geq 4$, it is obvious that $\mathfrak{H}(M)$ is reductive and $f = 0$ since $\mathfrak{H}(M) = \{0\}$.

Thus, the question to be considered in the two-dimensional case is whether M admits a K-E metric when $k \geq 3$. This problem was finally solved by Tian [37] through the efforts of Siu [35], Tian [36], Tian-Yau [38] et al. Their methods including that of Nadel [33] give sufficient conditions which are applicable to the higher-dimensional case but have no apparent relations with conditions such as the reductivity of $\mathfrak{H}(M)$ or the vanishing of f.

The next example is interesting when we compare Theorem 3.3 with the Lichnerowicz-Matsushima theorem.

EXAMPLE 3.7 ([15], [34]). Let m, n be positive integers, and let L_1, L_2 denote the hyperplane bundles of $\mathbb{P}^m(\mathbb{C})$, $\mathbb{P}^n(\mathbb{C})$, respectively. Let $E = \pi_1^* L_1 \oplus \pi_2^* L_2$ be a rank 2 complex vector bundle over $\mathbb{P}^m(\mathbb{C}) \times \mathbb{P}^n(\mathbb{C})$, where π_i denotes the ith ($i = 1, 2$) factor projection of $\mathbb{P}^m(\mathbb{C}) \times \mathbb{P}^n(\mathbb{C})$, and let M be the total space of the projective bundle $\mathbb{P}(E)$ of E. Then $c_1(M) > 0$ and $\mathfrak{H}(M)$ is reductive. Moreover, $f = 0$ if and only if $m = n$.

Considering the above example in the case of $m \neq n$, we can show that there exist examples of M such that $\mathfrak{H}(M)$ are reductive but $f \neq 0$ nevertheless. On the other hand, we know no examples of M such that the $\mathfrak{H}(M)$ are not reductive and $f = 0$. The necessary conditions for the existence of K-E metrics are satisfied when $m = n$, and it was proved by Sakane [34], Koiso-Sakane [25, 26] that there actually exist K-E metrics when $m = n$. Their method is as follows: the Einstein equation is reduced to an ordinary differential equation on the open subset

$$\mathbb{P}(E) \setminus \{\text{two sections } \{(1:0)\}, \{(0:1)\} \text{ of } \mathbb{P}(E)\}$$

of $\mathbb{P}(E)$ and the solution is smoothly extended to the two sections if $f = 0$. See also Mabuchi [31] about this result.

Finally, it should be added that the author got the idea for the character f from the integrability condition of Kazdan-Warner [24] for Nirenberg's problem.

§4. Some results concerning existence and uniqueness

In this section we give a short survey on the existence and uniqueness of K-E metrics and the Kähler metrics of constant scalar curvature (within the author's knowledge).

Let us begin with the survey of uniqueness. If g is an extremal Kähler metric, then for any automorphism a of M the pull-back a^*g is also an extremal Kähler metric. Let $A_0(M)$ denote the identity component of the group $A(M)$ consisting of automorphisms of M.

THEOREM 4.1 ([12]). *Fix the Kähler class Ω and let $\mathfrak{M} \subset \Omega^+$ denote the subspace of all extremal Kähler metrics. Then each connected component of $\mathfrak{M}$ is exactly an orbit of $A_0(M)$.*

In view of the above theorem, we are interested in the connectivity of $\mathfrak{M}$. This problem has been solved in the case of K-E metrics. When $c_1(M) < 0$ or $= 0$, uniqueness holds in the sense of Theorem 3.4. For $c_1(M) > 0$, we have the following lemma of Bando-Mabuchi.

THEOREM 4.2 [6]). *Assume that $c_1(M) > 0$ and set $\Omega = c_1(M)$. Then the subspace $\mathfrak{M} \subset \Omega^+$ of all K-E metrics is connected unless it is empty.*

Hence it follows that the K-E metric, if it exists, is unique up to homothety when $A(M)$ is discrete (and necessarily finite because $c_1(M) > 0$).

Secondly, we shall discuss existence. As was seen in §3, K-E metrics always exist when $c_1(M) < 0$ or when $c_1(M) = 0$. When $c_1(M) > 0$, there are some sufficient conditions obtained by Siu, Tian, Tian-Yau, Nadel et al. Their methods are different, and we shall outline Nadel's method in this section. Set $\Omega = c_1(M)$, fix any $g \in \Omega^+$, and define $F = F(g)$ by (3.5). Assume that a smooth function φ satisfies the complex Monge-Ampère equation:

$$(4.3) \qquad \frac{\det(g_{\alpha\bar{\beta}} + \varphi_{\alpha\bar{\beta}})}{\det(g_{\alpha\bar{\beta}})} = e^{-\varphi + F}.$$

Then we can see that $(g_{\alpha\bar{\beta}} + \varphi_{\alpha\bar{\beta}})$, namely, the deformation of the Kähler metric g

is a K-E metric by letting $\partial\overline{\partial}\log$ operate on both sides of (4.3). Thus, it suffices to solve (4.3). In order to solve (4.3), we consider the following equation for $0 \leq t \leq 1$:

$$(*(t)) \qquad \frac{\det(g_{\alpha\overline{\beta}} + \varphi_{\alpha\overline{\beta}})}{\det(g_{\alpha\overline{\beta}})} = e^{-t\varphi + F}.$$

Set $\mathscr{S} = \{t \in [0,1] \mid (*(t))$ can be solved$\}$. We shall prove that $\mathscr{S}$ is not empty, $\mathscr{S}$ is open, and $\mathscr{S}$ is closed. Then $1 \in \mathscr{S}$ and (4.3) can be solved. Now $(*(0))$ is actually the same equation that was solved in the proof of the existence of K-E metrics in the case $c_1(M) = 0$, and hence, $0 \in \mathscr{S} \neq \varnothing$. The openness of $\mathscr{S}$ was proved by Aubin [33] by using the implicit function theorem. Thus it remains only to show that $\mathscr{S}$ is closed. To do this, it is sufficient to show that, for $\{t_j\}_{j=1}^{\infty} \subset \mathscr{S}$ such that $t_j \to t_\infty$, the solution φ_j of $(*(t_j))$ has the estimate $\|\varphi_j\|_{C^3} \leq c$ for some constant c which is independent of j; because the above inequality leads to uniform boundedness and the equicontinuity of the kth derivatives ($k \leq 2$) of $\{\varphi_j\}$, which leads to the existence of C^2-convergent subsequence by using the Ascoli-Arzela theorem and, hence, to the existence of the solution φ_∞ of $(*(t_\infty))$. Using more refined arguments, we can show that $\varphi_\infty \in C^{2,\alpha}$ ($\alpha > 0$) and, further, that $\varphi_\infty \in C^\infty$ by use of the Schauder estimate.

Now, since it is shown in the paper of Yau [41] that the C^0-bounded sequence $\{\varphi_j\}$ is necessarily C^3-bounded, we have to do only the C^0-estimate. The proof is by contradiction. Assume that $\{\varphi_j\}$ is not C^0-bounded. Then using some estimates of $\{\varphi_j\}$, we can show that there exists an analytic subspace $V \in M$ that approximates the subset $\{p \in M \mid \sup_j |\varphi_j| = \infty\}$, where $(*(t_\infty))$ cannot be solved. Moreover, we can show that the above V has various properties such as
- (4.4) V is neither empty nor the whole of M,
- (4.5) $H^i(V, \mathscr{O}_V) = 0$ for $i \geq 1$,
- (4.6) V is a tree of smooth rational curves if $\dim V = 1$.

Assume that a compact Lie group G acts on M as automorphisms. If we set a Kähler metric g that is G-invariant, then all data is G-invariant, and hence, V is invariant under the action of the complexification $G_{\mathbb{C}}$ of G.

By using the above results, we see that Nadel's proof of the existence of K-E metrics on the three-point blow up M of $\mathbb{P}^2(\mathbb{C})$ is described as follows.

Let M be obtained by blowing up at $P_1 = (1:0:0)$, $P_2 = (0:1:0)$, $P_3 = (0:0:1)$. Let G be the compact Lie group generated by the following three kinds of automorphisms:
- (4.7) the permutations of homogeneous coordinates z_1, z_2, z_3,
- (4.8) $(z_1 : z_2 : z_3) \to (1/z_1 : 1/z_2 : 1/z_3)$,
- (4.9) $(z_1 : z_2 : z_3) \to (z_1 : e^{i\theta} z_2 : e^{i\theta} z_3)$ for $\theta \in \mathbb{R}$.

If M does not admit any K-E metric, then there exists a V that satisfies (4.4), (4.5), and (4.6). But the $G_{\mathbb{C}}$-invariant analytic subspace V, which is neither empty nor the whole of M, is the union of three exceptional curves and the proper transforms of the lines joining P_1 to P_2, P_2 to P_3, and P_3 to P_1 which has a 1-cycle. This contradicts (4.5), and hence, M admits a K-E metric.

Similarly to the above, the existence of K-E metrics on M is shown by contradiction by using the individual properties of each M.

Now we give a short survey on the existence of Kähler metrics of constant scalar curvature. There seem to be few general existence theorems for metrics of constant scalar curvature. This may stem from the following reasons: firstly, compared with the case of K-E metrics where the objects of interest are restricted to the manifolds with $c_1(M) > 0$, $c_1(M) = 0$, or $c_1(M) < 0$, the objects of interest are too numerous in the case of constant scalar curvature, and secondly, though the K-E metrics are described by elliptic partial differential equations of second order, the equation describing constant scalar curvature is of fourth order unless some contrivance is used.

Restricting the objects, we consider compact complex surfaces of zero scalar curvature. The problem of the classification of such surfaces was brought up in [42]. Here we introduce the result of Ito [23]. There are the following two necessary and sufficient conditions for the vanishing of the scalar curvature of a compact Kähler surface (M, g).

(4.10) (M, g) is antiselfdual. Namely, if $W = W_+ + W_-$ is the decomposition of the Weyl tensor W into the ± 1-eigenspaces of the Hodge operator $*$, then $W_+ = 0$.

(4.11) The first Chern form $\rho(g)$ is antiselfdual. Namely, $*\rho(g) = -\rho(g)$.

Now we have the following, where $P_1(M)$ denote the first Pontrjagin number:

$$P_1(M) = \frac{1}{4\pi^2} \int_M (|W_+|^2 - |W_-|^2) \, dV_g.$$

Hence, it follows from (4.10) that $P_1(M) \leq 0$. Moreover, it follows from (4.11) that

$$c_1^2(M) = \int_M \rho(g) \wedge \rho(g) = -\int_M \rho(g) \wedge *\rho(g) \leq 0.$$

Moreover, from the Weitzenböck formula for the sections of the canonical bundle K_M, we can see that the geometric genus $P_g = 0$ or K_M is trivial if the scalar curvature vanishes. Thus, using the classification theory, we have the following.

THEOREM 4.12 ([23]). *Let M be a compact Kähler surface of zero scalar curvature. Then M is one of the following*:
  (i) *a flat complex torus*;
  (ii) *a K3-surface or its quotient*;
  (iii) *a manifold obtained by blowing up $\mathbb{P}^2(\mathbb{C})$ more than ten times successively*;
  (iv) *a ruled surface of genus ≥ 2*;
  (v) *a manifold obtained by blowing up a ruled surface of genus 0 more than nine times successively*;
  (vi) *a manifold obtained by blowing up a ruled surface of genus 1 more than once successively.*

The above theorem does not assert that the surfaces of (i)–(vi) actually admit Kähler metrics of zero scalar curvature. Note that, when we consider the existence problem, we have to distinguish the problem of the existence with a fixed Kähler class from the problem of the existence with no restriction. If we consider the problem of the existence with a fixed Kähler class, there clearly exists an example of nonexistence in the case of (iv).

So we consider the problem of the existence with no restriction of the Kähler class. The scalar curvature for (i) and (iv) with respect to each standard metric is zero, and it follows from Yau's theorem that (ii) admits a Kähler metric of zero Ricci curvature, and hence, (ii) admits a Kähler metric of zero scalar curvature. Though there seem to be no known general results in the case of (iii), (v), and (vi), there is an interesting method by LeBrun [27] for constructing M in the special case of (vi). Roughly speaking, LeBrun's theorem is as follows.

THEOREM 4.13 ([27]). *Let S_g be a compact Riemann surface of genus ≥ 2. If $2k$ points $p_1, \ldots, p_k$ and $q_1, \ldots, q_k$ on $S_g \times \mathbb{P}^1(\mathbb{C})$ are in sufficiently symmetric position, then the complex surface M obtained by blowing up $S_g \times \mathbb{P}^1(\mathbb{C})$ at $p_1, \ldots, p_k, q_1, \ldots, q_k$ has a scalar-flat Kähler metric.*

The proof of the above theorem is as follows. Let X be the manifold obtained by removing $2k$ points that are pairwise symmetric with respect to $S_g \times \{0\}$ from the hyperbolic 3-manifold $S_g \times (-1, 1)$. Then it is shown that M is obtained by attaching $2k$ points and $S_g \times \{\pm 1\}$ to a certain S^1-bundle over X. Moreover, it is shown that the explicit metric on the total space of the S^1-bundle defined by using a Green function on X and a connection in the S^1-bundle has zero scalar curvature and that the metric is extended smoothly to the attached $2k$ points and $S_g \times \{\pm 1\}$ and defines a Kähler metric on M. Compared with the existence theorems of K-E metrics that are obtained by solving a partial differential equation, the above result is obtained by a constructive method.

§5. K-E metrics with $c_1 > 0$

Now we consider again the K-E metrics with $c_1(M) > 0$. As was stated in §§3 and 4, on the one hand, there are some results about the existence of such metrics, and, on the other hand, there are some obstructions to the existence of such metrics. Consider the following question.

QUESTION 5.1. Let M be a compact complex manifold with $c_1(M) > 0$. Then does M admit a K-E metric if $f = 0$?

Since $\mathfrak{H}(M) = 0$ clearly implies that $f = 0$, the next question is a special case of (5.1).

QUESTION 5.2. (cf. [42]). Let M be a compact complex manifold with $c_1(M) > 0$. Then does M admit a K-E metric if $\mathfrak{H}(M) = 0$?

Question (5.1) relates to the variational problem considered in §2. Let $\Omega = c_1(M)$ and let $g \in \Omega^*$ be an extremal Kähler metric. Then $\mathrm{grad}\,\sigma(g)$ is a holomorphic vector field. Moreover, if $f = 0$, then we have $\sigma(g) = m + \Delta F$ and hence, using integration by parts, we have

$$0 = f(\mathrm{grad}\,\sigma(g)) = \frac{(m+1)i}{2\pi} \int_M (\mathrm{grad}\,\sigma(g)) F \omega^m = -\frac{(m+1)i}{2\pi} \int_M (\Delta F)^2 \omega^m.$$

Therefore, we have $\Delta F = 0$, namely, F is equal to a constant. Hence $\rho_\omega = \omega$ and the extremal Kähler metric g is a K-E metric.

Thus, under the condition that $f = 0$, the existence of a K-E metric is equivalent to the existence of an extremal Kähler metric. What is known so far is that, as was stated in §2, there is an example of M and a Kähler class such that $f = 0$ and that M does not admit any extremal Kähler metric with the given Kähler class. But it is not known whether such an example exists if $c_1(M) > 0$ and $\Omega = c_1(M)$.

Now can we construct a counterexample of (5.2) if (5.2) is false? We shall consider the method stated below as a trial.

If the Lie algebra homomorphism $f: \mathfrak{H}(M) \to \mathbb{C}$ is lifted to a Lie group homomorphism, then it may become a nontrivial character of the Lie group of all automorphisms of M even when $\mathfrak{H}(M) = 0$. Does this lifted character become an obstruction to the existence of K-E metrics that is valid even when $\mathfrak{H}(M) = 0$? So far, though we have succeeded in lifting f to a group character, we do not know whether the lifted character is an obstruction to the existence of K-E metrics. We shall discuss this lifted character in the following sections.

§6. $GL(m; \mathbb{C})$-invariant polynomials and $A(M)$-invariant polynomials

Let M be a compact complex manifold with $c_1(M) > 0$. Set $\Omega = c_1(M)$, and define the smooth function F according to (3.5) for $g \in \Omega^+$. Now let us recall Yau's solution of Calabi's conjecture.

THEOREM 6.1 ([40]). *Let M be a compact Kähler manifold, and let $\tilde{\rho}$ be the real closed $(1,1)$-form that represents $c_1(M)$. Then there exists a unique Kähler metric g in each Kähler class such that $\rho(g) = \tilde{\rho}$.*

The above theorem implies that there exists a unique $h \in \Omega^+$ such that $\omega(g) = \rho(h)$. Hence (3.5) is equivalent to

$$\rho(g) - \rho(h) = \frac{i}{2\pi} \partial \bar{\partial} F,$$

and therefore we have

$$F = -\log \frac{\det g}{\det h}.$$

Using the above equality, we can rewrite the character f only in terms of h as follows.

(6.2)
$$\begin{aligned}
f(X) &= \frac{(m+1)i}{2\pi} \int_M X(F) \omega(g)^m \\
&= -\frac{(m+1)i}{2\pi} \int_M \frac{\det h}{\det g} X\left(\frac{\det g}{\det h}\right) \omega(g)^m \\
&= -\frac{(m+1)i}{2\pi} \int_M X\left(\frac{\det g}{\det h}\right) \omega(h)^m \\
&= \frac{(m+1)i}{2\pi} \int_M \mathrm{div}_h X \omega(g)^m \\
&= \frac{(m+1)i}{2\pi} \int_M \mathrm{div}_h X \rho(h)^m,
\end{aligned}$$

where $\mathrm{div}_h X = \sum_\alpha \nabla_\alpha X^\alpha$ is the divergence of X with respect to h. Note that the last term is defined for a Hermitian metric h, which is not necessarily a Kähler metric. Moreover, we have the following.

THEOREM 6.3 ([21]). *Let M be a compact complex manifold, and let h be a Hermitian metric of M. Denote $f: \mathfrak{H}(M) \to \mathbb{C}$ by the last term of (6.2). Then f is independent of the choice of h.*

At first, the author got the idea for f in the case where $c_1(M) > 0$, and later on, became aware that f belongs to each of the two different families: the family of Kähler invariants as in Theorem 3.3 and the family of invariants of complex manifolds as in Theorem 6.3.

For an $m \times m$ square matrix A, set

$$\det\left(1 + \frac{i}{2\pi}tA\right) = 1 + tc_1(A) + \cdots + t^m c_m(A).$$

The ring $I(\mathrm{GL}(m;\mathbb{C}))$ of $\mathrm{GL}(m;\mathbb{C})$-invariant polynomials is generated by $c_1, \ldots, c_m$. Since $c_1(A)$ is equal to $\frac{i}{2\pi}\operatorname{tr} A$, it follows from (6.2) that

(6.4) $$f(X) = \int_M c_1^{m+1}(L(X) + \Theta),$$

where $L(X) = \nabla_X - L_X \in C^\infty(M, \operatorname{End}(T'M))$ and Θ is the curvature form of h. Note that $\operatorname{tr} L(X) = \operatorname{div}_h X$ and $\frac{i}{2\pi}\operatorname{tr} \Theta = \rho(h)$.

More generally, we can see that

$$f_\varphi(X) = \int_M \varphi(L(X) + \Theta)$$

is independent of the choice of h for any $\mathrm{GL}(m;\mathbb{C})$-invariant polynomial φ of degree $m + k$. Hence, f_φ is invariant under the action of the automorphism group $A(M)$ and $\Psi: I(\mathrm{GL}(m;\mathbb{C})) \to I(A(M))$ is defined by $\Psi(\varphi) = f_\varphi$.

The above formulation reminds us of the classical results concerning the Chern classes. Actually, f_φ coincides with the Chern number when the degree of φ is equal to m. Furthermore, we have the following.

(6.5) Assume that the zero set $\operatorname{zero}(X)$ of X consists of isolated points and that $L(X)_p$ is nondegenerate for each $p \in \operatorname{zero}(X)$. Then we have

$$f_\varphi(X) = \sum_p \frac{\varphi(L(X)_p)}{\det L(X)_p}.$$

(6.5) is the same as the result about the Chern numbers (cf. [7]), and when $\operatorname{zero}(X)$ is a submanifold of M, $f_\varphi(X)$ can be calculated similarly as in [8] by using $\operatorname{zero}(X)$ and the characteristic classes of the normal bundle of $\operatorname{zero}(X)$.

The mapping Ψ can be regarded as integration along the fiber of an M-bundle in the following sense.

THEOREM 6.6 ([21]). *Let $E_{\mathrm{GL}} \to B_{\mathrm{GL}}$ be the universal $\mathrm{GL}(m;\mathbb{C})$-bundle and $E_{A(M)} \to B_{A(M)}$ the universal $A(M)$-bundle. Let FM be the frame bundle of M. Then $FM_{A(M)} = E_{A(M)} \times_{A(M)} FM$ is the frame bundle along the fiber of $\pi: M_{A(M)} = E_{A(M)} \times_{A(M)} M \to B_{A(M)}$, and the following diagram is commutative:*

$$\begin{array}{ccc} I^{m+k}(\mathrm{GL}(m;\mathbb{C})) & \xrightarrow{\Psi} & I^k(A(M)) \\ \downarrow W & & \downarrow W \\ H^{m+k}(M_{A(M)}) & \xrightarrow{\pi_*} & H^k(B_{A(M)}) \end{array}$$

where π_* is the Gysin homomorphism (i.e., integration along the fiber), the left W is the Weil homomorphism of the $\mathrm{GL}(m;\mathbb{C})$-bundle $FM_{A(M)} \to M_{A(M)}$, and the right W is the Weil homomorphism of the $A(M)$-bundle $E_{A(M)} \to B_{A(M)}$.

We have the following as a corollary of (6.6).

(6.7) $f(X)$ is an integer if $\exp X = 1 \in A(M)$.

§7. Lifting to a group homomorphism

Let $A^\delta(M)$ be the automorphism group of M with the discrete topology, and let $E_{A^\delta(M)} \to B_{A^\delta(M)}$ be the universal $A^\delta(M)$-bundle. Then $M_{A^\delta(M)} = E_{A^\delta(M)} \times_{A^\delta(M)} M \to B_{A^\delta(M)}$ is a flat M-bundle, and hence, its horizontal subspaces define a foliation $\mathscr{F}$ of real codimension $2m$. Moreover, since M is a complex manifold, the transversal direction to the leaves of $\mathscr{F}$ has a holomorphic structure. A foliation such as this is called the transversally holomorphic foliation. The normal bundle $\nu(\mathscr{F}) \to M_{A^\delta(M)}$ naturally has the structure of a complex vector bundle. A connected θ of $\nu(\mathscr{F})$ is called a basic connection if θ is orthogonal to $\mathscr{F}$, namely, $\theta|_{T\mathscr{F}} = 0$ and $\mathscr{O}$ is of type $(1,0)$ with respect to the complex structure of the transversal direction. Since the curvature form $\Theta = d\theta + \theta \wedge \theta$ of a basic connection θ has a $(1,0)$-component of the transversal direction, we have $\varphi(\Theta) = 0$ for any $\varphi \in I^{m+k}(\mathrm{GL}(m; \mathbb{C}))$ $(k > 0)$. (This follows from the Bott vanishing theorem for the transversally holomorphic foliation [9].) Hence, we can use the Chern-Simons theory to define

$$S_\varphi(\mathscr{F}) \in H^{2m+2k-1}(M_{A^\delta(M)}; \mathbb{C}/\mathbb{Z}).$$

It is known that $S_\varphi(\mathscr{F})$ is independent of the choice of basic connections θ ([13]). In particular, when $k = 1$, using the Gysin homomorphism, we have

$$\pi_* S_\varphi(\mathscr{F}) \in H^1(M_{A^\delta(M)}; \mathbb{C}/\mathbb{Z}) \cong \mathrm{Hom}(A^\delta(M), \mathbb{C}/\mathbb{Z}).$$

Furthermore, if we set $\varphi = c_1^{m+1}$, $\hat{f} = -\pi_* S_\varphi(\mathscr{F})$ becomes a lifting of f to the group $A^\delta(M)$.

Now we wish to know whether or not the existence of K-E metrics implies $\hat{f} = 0$. In order to settle the question, we have to rewrite $\hat{f}$ in more explicit form. Here the imaginary part of $\hat{f}$, $\mathrm{Im}\,\hat{f}: A(M) \to \mathbb{R}$, can be expressed as follows.

THEOREM 7.1 ([17]). *For $a \in A(M)$, we have*

$$\mathrm{Im}\,\hat{f}(a) = \frac{m+1}{2\pi} \int_M \log \frac{\det a^* h}{\det h} \sum_{k=0}^m a^* \rho(h)^k \wedge \rho(h)^{m-k},$$

where h is an arbitrary Hermitian metric.

We can show that $\mathrm{Im}\,\hat{f} = 0$ if M admits a K-E metric. Actually, if $c_1(M) < 0$ or $c_1(M) = 0$, then it follows immediately from the uniqueness of K-E metrics that $\mathrm{Im}\,\hat{f} = 0$. If $c_1(M) > 0$ and M admits a K-E metric, then $f = 0$ and $\mathrm{Im}\,\hat{f}$ defines a homomorphism from $A(M)/A_0(M)$ to $\mathbb{R}$. But since $c_1(M) > 0$, $A(M)/A_0(M)$ is a finite group, and hence, $\mathrm{Im}\,\hat{f} = 0$.

There seems to be no integration formula for the real part $\mathrm{Re}\,\hat{f}: A(M) \to \mathbb{R}/\mathbb{Z}$ of $\hat{f}$ such as (7.1). In the next section, according to [22, 39], when $a \in A(M)$ has a finite order, we shall see that $\mathrm{Re}\,\hat{f}(a) = \hat{f}(a)$ can be interpreted as the eta invariant and can be expressed in terms of the fixed point set of a and the characteristic classes of its normal bundle.

The character $\mathrm{Im}\,\hat{f}$ was also obtained by Mabuchi [30] in the case where $c_1(M) > 0$. Actually, the cocycle condition of the K-energy map in [30] is equivalent to the condition that $\mathrm{Im}\,\hat{f}$ is a character. It is known that there exists an "almost K-E metric" if the K-energy map is bounded from below ([4]).

§8. Interpretation as an eta invariant

Let M be a compact complex manifold. Let a be an automorphism of M whose order is n, and let K be the cyclic group generated by a. Define the action of K on $M \times S^1$ by $a(p, e^{i\theta}) = (a(p), e^{i(\theta + 2\pi/n)})$. Let $M_a = (M \times S^1)/K$ be the mapping torus defined by a, and let

$$\xi = \left(\pi^* \bigotimes^{m+1} (K_M^{-1} - \varepsilon^1) \right) \Big/ K$$

be the virtual vector bundle defined by the anticanonical bundle K_M^{-1}, trivial complex line bundle ε^1, and the natural projection $\pi \colon M \times S^1 \to M$. A K-invariant Hermitian metric on M defines natural Hermitian metrics on M_a and ξ. Let

$$A_\xi \colon \Gamma \left(\bigwedge^{\text{even}} T^* M_a \otimes \xi \right) \to \Gamma \left(\bigwedge^{\text{even}} T^* M_a \otimes \xi \right)$$

be the first-order selfadjoint differential operator defined by

$$A_\xi \phi = i^{m+1}(-1)^{q+1}(*d_\xi - d_\xi *)\phi \quad \text{for } \phi \in \Gamma \left(\bigwedge^{2q} T^* M_a \otimes \xi \right),$$

where d_ξ is the covariant exterior differential operator of ξ.

THEOREM 8.1 ([22]). *Let η_ξ be the eta invariant of A_ξ. Then we have*

$$2^{m+1} \hat{f}(a) = \eta_\xi \bmod \mathbb{Z}.$$

This theorem is proved as follows. Let W be a $(2m+2)$-dimensional almost complex manifold such that $\partial W = M$ and extend ξ to W. Then it follows from the Atiyah-Patodi-Singer theorem ([1]) for the signature operator $D_\xi = d_\xi + d_\xi *$ on W that

$$\eta_\xi = 2^{m+1} \int_W \text{ch}(\xi) \mathscr{L}(W) - \text{sign}(W, \xi)$$

$$= 2^{m+1} \int_W c_1(TW)^{m+1} \bmod \mathbb{Z}.$$

On the other hand, it follows from the property of the Chern-Simons invariants that

$$S_{c_1^{m+1}}(\mathscr{F})[M_a] = \int_W c_1(TW)^{m+1} \bmod \mathbb{Z}.$$

Thus we can complete the proof of (8.1).

Now consider the signature operator on $M \times D^2$ whose restriction to the boundary $M \times S^1$ defines A_ξ similarly as above. Let $\eta_\xi(a^k)$ denote the equivariant eta invariant of A_ξ for $a^k \in K$. Then η_ξ in (8.1) can be written as follows:

$$\eta_\xi = \frac{1}{n} \sum_{k=0}^{n-1} \eta_\xi(a^k).$$

$\eta_\xi(a^k)$ can be expressed in terms of the fixed point set of a^k in $M \times D^2$ and the characteristic classes of its normal bundle ([14]). Furthermore, since the fixed point set of a^k in $M \times D^2$ coincides with the fixed point set of a^k in M, $\eta_\xi(a^k)$ can be expressed in terms of the fixed point set in M and the characteristic classes of its normal bundle. For detail, see [**22**].

Recently, using the Spin^C-Dirac operator, Tsuboi [**39**] improved (8.1) so as to remove 2^{m+1} from the left-hand term of (8.1).

THEOREM 8.2 ([**39**]). *Let A_ξ be the Spin^C-Dirac operator on M_a defined similarly as in* (8.1). *Then we have*

$$\hat{f}(a) = \frac{1}{2}\eta_\xi \bmod \mathbb{Z}.$$

Using (8.2), we can see that $\hat{f}(a) = 0$ if $c_1(M) > 0$, $m = 2$, and $\mathrm{order}(a) = 2$.

Let $A_\xi^+ : \Gamma(S^+ \otimes \xi) \to \Gamma(S^- \otimes \xi)$ be the Spin^C-Dirac operator defined by the natural $\mathrm{Spin}^C(2m)$-structure of M. Comparing the expression of η_ξ in terms of the fixed point set with the Atiyah-Singer equivariant index theorem for A_ξ^+, we have the following.

THEOREM 8.3 ([**39**]).

$$e^{2\pi i \hat{f}(a)} = \det(a| \operatorname{Ker} A_\xi^+)^* \det(a| \operatorname{Ker}(A_\xi^+)^*).$$

The above equality is equivalent to Witten's holonomy formula for a locally flat M-bundle. Moreover, the equivariant index of the Spin^C-Dirac operators in the above case is equal to the equivariant index of Dolbeault operators, and hence (8.3) leads to the result of [**19**].

References

1. M. F. Atiyah, V. K. Patodi, and I. M. Singer, *Spectral asymmetry and Riemannian geometry.* I, Math. Proc. Cambridge Philos. Soc. **77** (1975), 43–69.
2. T. Aubin, *Equations du type Monge-Ampère sur les variétés Kähleriennes compactes*, C. R. Acad. Sci. Paris **283** (1976), 11–121.
3. _____, *Réduction du cas positif de l'équation de Monge-Ampère sur les variétés Kähleriennes compactes à la démonstration d'une inégalité*, J. Funct. Anal. **57** (1984), 143–153.
4. S. Bando, *The K-energy map, almost Einstein-Kähler metrics and an inequality of the Miyaoka-Yau type*, Tôhoku Math. J. **39** (1987), 231–235.
5. S. Bando and T. Mabuchi, *On some integral invariants on complex manifolds.* I, Proc. Japan Acad. **62** (1986), 197–200.
6. _____, *Uniqueness of Einstein-Kähler metrics modulo connected group actions*, Algebraic Geometry, Sendai, 1985, Adv. Stud. Pure Math., vol. 10, Kinokuniya, Tokyo, 1987, pp. 11–50.
7. R. Bott, *Vector fields and characteristic numbers*, Michigan Math. J. **14** (1967), 231–244.
8. _____, *A residue formula for holomorphic vector fields*, J. Differential Geom. **1** (1967), 311–330.
9. _____, *On a topological obstruction to integrability*, Global Analysis (S.-S. Chern and S. Smale, eds.), Proc. Sympos. Pure Math., vol. 16, Amer. Math. Soc., Providence, R.I., 1970, pp. 127–131.
10. D. Burns and P. de Bartolomeis, *Stability of vector bundles and extremal metrics*, Invent. Math. **92** (1988), 403–407.
11. E. Calabi, *Extremal Kähler metrics*, Seminar on Differential Geometry (S.-T. Yau, ed.) Ann. of Math. Stud., vol. 102, Princeton Univ. Press, Princeton, N.J., 1982, pp. 259–290.
12. _____, *Extremal Kähler metrics.* II, Differential Geometry and Complex Analysis (I. Chaval and H. M. Farkas, eds.), Springer, Berlin, 1985, pp. 95–114.
13. J. Cheeger and J. Simons, *Differential characters and geometric invariants*, Geometry and Topology (J. Alexander and J. Harer, eds.), Lecture Notes in Math., vol. 1167, Springer, Berlin, 1985, pp. 50–80.
14. H. Donnelly, *Eta invariants for G-spaces*, Indiana Univ. Math. J. **27** (1978), 889–918.
15. A. Futaki, *An obstruction to the existence of Einstein-Kähler metrics*, Invent. Math. **73** (1983), 437–443.
16. _____, *On compact Kähler manifolds of constant scalar curvature*, Proc. Japan Acad. **59** (1983), 401–402.

17. _____, *On a character of the automorphism group of a compact complex manifold*, Invent. Math. **87** (1987), 655–660.
18. _____, *Kähler-Einstein metrics and integral invariants*, Lecture Notes in Math., vol. 1314, Springer, Berlin, 1988.
19. A. Futaki and T. Mabuchi, *An obstruction class and a representation of holomorphic automorphisms*, Geometry and Analysis on Manifolds (T. Sunada, ed.), Lecture Notes in Math., vol. 1339, Springer, Berlin, 1988, pp. 127–140.
20. A. Futaki, T. Mabuchi, and Y. Sakane, *Einstein-Kähler manifolds with positive Ricci curvature*, Kähler metrics and moduli spaces (T. Ochiai, ed.), Adv. Stud. Pure Math., vol. 18-II, Kinokuniya, Tokyo, 1990, pp. 11–83.
21. A. Futaki and S. Morita, *Invariant polynomials of the automorphism group of a compact complex manifold*, J. Differential Geom. **21** (1985), 135–142.
22. A. Futaki and K. Tsuboi, *Eta invariants and automorphisms of compact complex manifolds*, Recent Topics in Differential and Analytic Geometry (T. Ochiai, ed.), Adv. Stud. Pure Math., vol. 18-I, Kinokuniya, Tokyo, 1990, pp. 251–270.
23. M. Ito, *Self-duality of Kähler surfaces*, Compositio Math. **51** (1984), 265–273.
24. J. Kazdan and F. Warner, *Curvature functions for compact 2-manifolds*, Ann. of Math. **99** (1974), 14–47.
25. N. Koiso and Y. Sakane, *Non-homogeneous Kähler-Einstein metrics on compact complex manifolds*, Curvature and Topology of Riemannian Manifolds, Lecture Notes in Math., vol. 1201, Springer, Berlin, 1985, pp. 165–179.
26. _____, *Non-homogeneous Kähler-Einstein metrics on compact complex manifolds*. II, Osaka J. Math. **25** (1988), 933–959.
27. C. LeBrun, *Scalar-flat Kähler metrics on blow-up ruled surfaces*, preprint.
28. M. Levine, *A remark on extremal Kähler metrics*, J. Differential Geom. **21** (1985), 73–77.
29. A. Lichnerowicz, *Géométrie des groupes de transformations*, Dunod, Paris, 1958.
30. T. Mabuchi, *K-energy maps integrating Futaki invariants*, Tôhoku Math. J. **38** (1986), 575–593.
31. _____, *Einstein-Kähler forms, Futaki invariants and convex geometry on toric Fano varieties*, Osaka J. Math. **24** (1987), 705–737.
32. Y. Matsushima, *Sur la structure du groupe d'homéomorphismes analytiques d'une certain variété Kählerienne*, Nagoya Math. J. **11** (1957), 145–150.
33. A. M. Nadel, *Multiplier ideal sheaves and Kähler-Einstein metrics of positive scalar curvature*, Ann. of Math. **132** (1990), 549–596.
34. Y. Sakane, *Examples of compact Kähler-Einstein manifolds with positive Ricci curvature*, Osaka J. Math. **31** (1986), 585–617.
35. Y.-T. Siu, *The existence of Kähler-Einstein metrics with positive anticanonical line bundle and a suitable finite symmetry group*, Ann. of Math. **127** (1988), 585–627.
36. G. Tian, *On Kähler-Einstein metrics on certain Kähler manifolds with $c_1(M) > 0$*, Invent. Math. **89** (1987), 225–246.
37. _____, *On Calabi's conjecture for complex surfaces with positive first Chern class*, Invent. Math. **101** (1990), 101–172.
38. G. Tian and S.-T. Yau, *Kähler-Einstein metrics on complex surfaces with $c_1(M) > 0$*, Comm. Math. Phys. **112** (1987), 175–203.
39. K. Tsuboi, *The lifted Futaki invariants and the index of Spin^C-Dirac operators*, manuscript.
40. S.-T. Yau, *On Calabi's conjecture and some new results in algebraic geometry*, Proc. Nat. Acad. Sci. USA **74** (1977), 1798–1799.
41. _____, *On the Ricci curvature of a compact complex manifold and the complex Monge-Ampère equation*. I, Comm. Pure Appl. Math. **31** (1978), 339–441.
42. _____, *Problem section in seminar on differential geometry*, (S.-T. Yau, ed.), Ann. of Math. Stud., vol. 102, Princeton Univ. Press, Princeton, N.J., 1982, pp. 669–706.

Tokyo Institute of Technology, Oh-okayama, Meguro-ku, Tokyo 152, Japan

Translated by Kenji Tsuboi

A Convergence Theorem for Einstein Metrics and the ALE Spaces

Hiraku Nakajima

Introduction

A Riemannian metric g on a manifold is called an Einstein metric if its Ricci curvature is a constant multiple of g, i.e., $\mathrm{Ric} = kg$ for some constant k. Although this is supposed to be an important equation for space-time in relativity, it seems difficult to understand the meaning of this definition itself at first sight. We recognize its importance only by the experience of seeing that Einstein metrics have appeared in many areas of differential geometry. For example, if you read the book by Besse [Be], you realize that so many fields relate to Einstein metrics. One of the most important results is the existence of the Einstein-Kähler metric obtained by Aubin [Au] and Yau [Ya] (shown by Yau when $c_1 = 0$, and by Aubin and Yau independently when $c_1 < 0$). This result had many applications to algebraic geometry. Recently in works by Narashimhan-Seshadri [NaSe], S. Kobayashi [Kos], Lübke [Lü], Donaldson [Do1, 2], Uhlenbeck-Yau [UY], and others, it was pointed out that an Einstein-Hermitian metric, which is a counterpart of the Einstein-Kähler metric in gauge theory, has a close relationship with stable holomorphic vector bundles on Kähler manifolds. By their results it becomes possible to study the moduli space of holomorphic vector bundles from the differential geometric viewpoint. (See the textbook by Siu [Si].)

It is believed that Einstein-Kähler metrics also relate to the moduli space of complex structures. Then it is natural to consider its compactification. Anderson [An1] and Bando-Kasue-Nakajima [BaKaNa][1] obtained a convergence theorem for Einstein metrics when the manifold is four-dimensional. We used the theory of Hausdorff distance, which was introduced by Gromov and developed by many people. (It should be noticed that our result is motivated by the study of degeneration of K3 surfaces by R. Kobayashi-Todorov [KorTo].) In particular, we obtain a compactification of the moduli space when the Ricci curvature is positive, and a completion when nonpositive. As a limit we get a space called an orbifold, which is not a manifold.

From the analytic point of view regarding the Einstein metric as a solution of a PDE, one can find many cousins of the convergence theorem in other nonlinear PDEs of elliptic type which have "scaling invariance". In these problems, the Palais-Smale condition (C) does not hold; that is, there may not exist a convergent subsequence in a given sequence of solutions with bounded energy. But one can show that solutions

---

1991 *Mathematics Subject Classification.* Primary 53C55.
This article orginally appeared in Japanese in Sūgaku **44** (2) (1992), 133–146.

[1] This is a joke by B&O. "Bakana" means foolish in Japanese !

converge outside a finite set thanks to the scaling invariance. One can find examples in Sacks-Uhlenbeck [**SaUh**], Uhlenbeck [**Uh**], Sedlacek [**Se**], Brezis-Coron [**BrCo**], Bahri-Coron [**BaCo**], and so on.

As a by-product of the convergence theorem, there appears naturally a family of noncompact Ricci-flat manifolds, called ALE spaces. (It corresponds to a "bubble" in the case of harmonic maps or Yang-Mills connections.) The ALE stands for *asymptotically locally Euclidean* and means that the metric approximates the standard metric on $\mathbb{R}^4/\Gamma$ at infinity. The ALE spaces with hyper-Kähler structures already appeared in a very different context. Physicists, Eguchi-Hanson [**EgHa**] and Gibbons-Hawking [**GiHa**] constructed ALE spaces corresponding to cyclic groups. On the other hand, Hitchin [**Hi**] constructed the same space by using the twistor method. His construction suggested to us a relationship to deformation theory of the simple singularity $\mathbb{C}^2/\Gamma$. Thus he conjectured that there exist similar spaces corresponding to other finite subgroups of $\mathrm{SU}(2)$ (i.e., the binary polyhedral groups). This conjecture was solved affirmatively by Kronheimer using the method called hyper-Kähler quotients. His construction gives a hyper-Kähler structure on each fiber of the semi-universal deformations and their simultaneous resolutions, which were constructed by Brieskorn [**Br**] and Slodowy [**Sl**] by using nilpotent varieties of Lie algebras.

ALE spaces with hyper-Kähler structures can be considered as analogues of K3 surfaces in several points. The author [**Na2**] studied the moduli space of anti-self-dual connections motivated by the study of stable holomorphic vector bundles over K3 surfaces by Mukai [**Mu**]. The moduli space has a hyper-Kähler structure and becomes again an ALE space when its dimension is four. But the base space and the moduli space are not homeomorphic to each other in general, in contrast with the case of K3 surfaces. This result was further developed by Kronheimer-Nakajima [**KrNa**]. We gave an analogue of the ADHM construction for the anti-self-dual connections on S^4. Our theory relates to the representations of quivers on the extended Dynkin diagrams. All these results show us the richness of the geometry of ALE spaces.

In this article we shall explain the above results, but we cannot go into moduli spaces of anti-self-dual connections because of limitation of space. I hope to do so on another occasion. In §1, we shall introduce the convergence theorem. In §2, we shall see how ALE spaces bubble out. Then in §3, we shall explain the construction of ALE spaces with hyper-Kähler structures. §3 can be read independently of the previous sections, although §2 depends on §1. The statements of the theorems are analytic, but we must use the language of algebraic geometry to give examples. The reader may wish to read only the part in which he/she has an interest.

§1. A convergence theorem for Einstein metrics

As we said in the introduction, an Einstein metric is a solution of a PDE describing space-time. Here we consider a sequence of Einstein metrics and study its convergence. This means that we study the situation when space-time changes and breaks down. Such convergence theorems appear very often in the study of elliptic partial differential equations, and actually our convergence theorem is strongly motivated by Sacks-Uhlenbeck's theorem on harmonic maps [**SaUh**] and Uhlenbeck's compactness theorem for Yang-Mills connections [**Uh1, 2**]. So we first recall the case of harmonic maps, which seems simplest, then proceed to the case of Einstein metrics.

THEOREM 1.1 ([**SaUh**]). *Let (X, g) be a compact two-dimensional Riemannian manifold, and let (M, h) be a compact Riemannian manifold of an arbitrary dimension.*

Let $\{f_i: X \to M\}$ be a sequence of harmonic maps with

$$E(f_i) \stackrel{\text{def.}}{=} \int_X |df_i|^2 dV_g \leq E < \infty,$$

where E is a positive constant independent of i. Then there exists a subsequence $\{j\} \subset \{i\}$ satisfying the following properties:
  (1) There exists a finite set $S = \{x_1, \ldots, x_k\}$ in X such that f_j converges to a map f_∞ in $C^\infty_{\text{loc}}(X \setminus S)$.
  (2) The limit map $f_\infty: X \setminus S \to M$ extends smoothly to the whole space X.

The second statement is deduced from the following removable singularities theorem.

THEOREM 1.2 ([SaUh]). *Let D be the 2-disk. If a harmonic map f defined over $D \setminus \{0\}$ satisfies*

$$\int_{D \setminus \{0\}} |df|^2 < \infty,$$

it extends smoothly across the singularity.

We give an example of convergence. Take the Riemann sphere $S^2 = \mathbb{C} \cup \{\infty\}$ for both (X, g) and (M, h), and consider the sequence of rational functions given by $\{f_i(z) = iz\}$.[2] The energy $E(f_i)$ is independent of i. When i goes to infinity, $f_i(z)$ converges to ∞ if $z \neq 0$, but f_i does not converge uniformly in a neighbourhood of 0. The limit is the constant map, which, of course, extends to the whole space S^2.

If we consider the energy density $|df_j|^2 dV$ as a measure on X in Theorem 1.1, it converges to

$$|df_\infty|^2 dV + \sum_{x \in S} a_x \delta_x,$$

where δ_x is the Dirac measure at x and a_x is a positive constant. That is to say, the energy density becomes concentrated around S and goes to the Dirac measure in the limit. (Remark: We have a similar result when the dimension of X is greater than 2. But the singular set S becomes $(n-2)$-dimensional, where $n = \dim X$.)

When the target manifold is isometrically embedded in the Euclidean space, the harmonic map equation can be written as $\Delta f + \Pi(df, df) = 0$, where Π is the second fundamental form. It is a nonlinear elliptic PDE of second order and the Euler-Lagrange equation for the energy functional. When we study analytic aspects of the harmonic map equation, it is natural to introduce the function space[3] $W^{1,2}$. It is quite essential in the proof of Theorem 1.1, especially for the finiteness of the singular set, that $p = 2$ is a critical exponent in the two-dimensional case. (Here we say p is a critical exponent, if any function in the Sobolev space $W^{1,q}$ is continuous if $q > p$, but not continuous when $q = p$.) It is also important that the harmonic map equation has "scaling invariance" in two dimensions. For an Einstein metric g

---

[2] A holomorphic map between Kähler manifolds is harmonic. In fact, it gives a minimum of the energy in its homotopy class.

[3]
$$W^{1,p} \stackrel{\text{def.}}{=} \{f \mid \int_X |f|^p dV + \int_X |df|^p dV < \infty\}$$

on a 4-manifold X, the counterpart to the energy for the harmonic map is the square of the L^2-norm of the full curvature tensor

$$\int_X |R_g|^2 dV_g.$$

(In fact, the Einstein metric gives the minimum of this functional defined on the set of metrics on X.) The curvature tensor is given by the second-order differential of the metric, so one must use $W^{2,2}$ as the function space. In the 4-dimensional case, the exponent $p = 2$ has a similar property to the above concerning $W^{2,q}$. So it is natural to expect a similar convergence theorem. Actually we have the following:

THEOREM 1.3 ([**An1, Na1, BaKaNa**]). *Let $\{(X_i, g_i)\}$ be a sequence of pairs of 4-dimensional compact manifolds and Einstein metrics with*

$$\operatorname{Ric} g_i = k g_i, \qquad \operatorname{diam}(X_i, g_i) \leq D,$$
$$\operatorname{vol}(X_i, g_i) \geq V, \qquad \textit{the Euler number of } X_i \leq R,$$

where k, D, V, R are constants independent of i. We also assume that k is normalized to be 0 or ± 1. Then there exists a subsequence $\{j\} \subset \{i\}$ with the following properties:

(1) $\{(X_j, g_j)\}$ *converges to a compact metric space X_∞ in the following sense:*[4] *If we remove a finite set $S = \{x_1, \ldots, x_k\}$ from X_∞, a C^∞-manifold structure and an Einstein metric g_∞ are defined over there. There exists a diffeomorphism F_j from $X_\infty \setminus S$ into X_j such that $F_j^* g_j$ converges to g_∞ in the $C^\infty_{\text{loc}}(X_\infty \setminus S)$-topology.*

(2) *The manifold structure and the Einstein metric g_∞ on $X_\infty \setminus S$ extend to the whole of X as an orbifold structure and an orbifold Einstein metric.*

By an orbifold structure and an orbifold metric we mean the following:

(a) $X_\infty \setminus S$ is a C^∞-manifold and the restriction of g_∞ is a Riemannian metric.

(b) For each singular point x_n, there exists a neighbourhood N_n such that $N_n \setminus \{x_n\}$ is diffeomorphic to $B^4 \setminus \{0\}/\Gamma$, where B^4 is a 4-dimensional unit ball and Γ is a finite subgroup of $O(4)$ acting freely on $B^4 \setminus \{0\}$. If we pull back the metric g_∞ to $B^4 \setminus \{0\}$, it extends smoothly across the singular point 0. (Γ may depend on the singular point x_n.)

As in the case of harmonic maps, we have the removable singularities theorem, but we omit the statement because the conditions are complicated.

The same result under an additional assumption on the lower bound of the injectivity radius was independently obtained by Gao [**Ga**]. In this case, S is an empty set, and situations like Example 1.5 given later do not appear.

REMARK 1.4. Since we do not give the proof of the theorem at all, the reader may have difficulty understanding the meaning of the conditions. The conditions are used to derive the estimates of the isoperimetric constants and are indispensable for obtaining the *a priori* estimates for curvature. From the point of view of the user of the theorem, the conditions are quite natural as:

(1) The assumption k to be 0 or ± 1 is always satisfied if you multiply the metric by an appropriate positive constant. So this condition is not essential.

(2) When $k = 1$, the condition $\operatorname{diam}(X_i, g_i) \leq D$ follows from Myers' theorem.

---

[4] In fact, convergence is with respect to Hausdorff distance introduced by Gromov. We do not talk about Hausdorff distance here, but note that the recent development in the theory of Hausdorff convergence is quite essential in the proof of Theorem 1.3.

(3) For a 4-manifold admitting an Einstein metric, the Euler number is given by the (universal) constant multiple of the square integral of the curvature tensor.

(4) The assumption dim $X_i = 4$ does *not* come from the fact that the dimension[5] of space-time is equal to four. It comes from a technical reason. In higher dimensions, the author conjectures that the sequence converges outside a singular set of codimension four if the L^2-norm of the curvature is uniformly bounded.

As in the harmonic map case, the measure $|R_{g_j}|^2 dV_{g_j}$ converges to

$$|R_{g_\infty}|^2 dV_{g_\infty} + \sum_{x \in S} a_x \delta_x$$

in a certain sense. (Since the spaces are changing, the usual convergence does not make sense.) The second term will be explained in the next section. Recall that the curvature measures how the space curves. As $j \to \infty$, the spaces bend around S and become singularities in the limit. But the singularities are quite simple as finite quotients of manifolds. So it is nothing fearful!

We now give examples.

EXAMPLE 1.5 ([**Pa, KorTo**]). Let X_∞ be an orbifold given by the $\mathbb{Z}/2\mathbb{Z}$-quotient of the complex 2-torus $T = \mathbb{C}^2/\mathbb{Z}^4$, where $-1 \in \mathbb{Z}/2\mathbb{Z}$ acts on T by

$$(z_1, z_2) \mod \mathbb{Z}^4 \mapsto (-z_1, -z_2) \mod \mathbb{Z}^4.$$

The flat metric on T descends to an orbifold metric g_∞ on X_∞, and it is an orbifold Einstein metric. Moreover, X_∞ has a complex manifold structure (with singularities) since the $\mathbb{Z}/2\mathbb{Z}$-action is holomorphic. Let us take the minimal resolution $\pi \colon X \to X_\infty$. The singularities are sixteen simple singularities of type A_1. The minimal resolution X is called a Kummer surface and is an example of K3 surfaces. Let $S = \{x_1, \ldots, x_{16}\}$ be the singular set, and let $E_1, \ldots, E_{16}$ be the exceptional sets. These are complex submanifolds of X biholomorphic to $\mathbb{C}P^1$ with the self-intersection number -2. By the solution of the Calabi conjecture by Yau [**Ya**] we have a unique Ricci-flat Kähler metric in each Kähler class.[6] Take a Kähler class, and thus a Ricci-flat Kähler metric g_i, as follows:

(1) the volume of X with respect to g_i is equal to 1;

(2) the volume of the exceptional set E_n is equal to $1/i$ for $n = 1, \ldots, 16$.

It can be shown that the metric g_i converges to $\pi^* g_\infty$ over $X \setminus \bigcup E_n$, but condition (2) forces the metric to become degenerate along E_n as $i \to \infty$. Since $X \setminus E_n$ is diffeomorphic to $X_\infty \setminus S$ by the map π, this gives an example of Theorem 1.3. Note that the limit object is (X_∞, g_∞), not $\pi^* g_\infty$. The limit metric $\pi^* g_\infty$ is degenerate along E_n, and the distance between two points in E_n becomes 0. Hence it is more natural to collapse E_n to a point! (See Figure 1.1 on p. 84.)

The moduli space of polarized K3 surfaces (i.e., pairs of K3 surfaces and Kähler classes with the volume 1) is known to be an open dense subset $K\Omega$ of

$$\overline{K\Omega} = \Gamma \backslash SO_o(3, 19)/SO(2) \times SO(19),$$

where Γ is the group of automorphisms preserving the intersection form on $H^2(X; \mathbb{Z})$. If we add singular K3 surfaces as above to the moduli space, we get the whole space

---

[5] I mean the dimension of space-time in relativity, not in superstring theory.
[6] An Einstein metric with Ric $= 0$ is called a Ricci-flat metric.

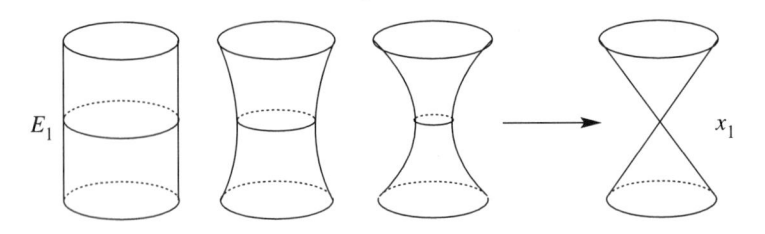

FIGURE 1.1. The behaviour of the metrics around the singular point x_1

$\overline{K\Omega}$. Thus Theorem 1.3 gives a natural completion of the moduli space (see [**KorTo, An2**] for details).

Similarly, by using the existence of the Einstein-Kähler metric[7] obtained by Aubin and Yau, one can also give examples with $k = -1$. Consider a deformation of projective surfaces of general type $\pi \colon \mathscr{X} \to \Delta$ such that $X_t = \pi^{-1}(t)$ satisfies $c_1(X_t) < 0$ (namely, the canonical bundle is ample) when $t \neq 0$, and the central fiber X_0 satisfies $c_1(X_0) \leq 0$ (the canonical bundle is nef), but $c_1(X_0) \not< 0$. When $t \neq 0$, X_t has a unique Einstein-Kähler metric, but X_0 does not. The limit of the Einstein-Kähler metric as $t \to 0$ is the orbifold Einstein-Kähler metric on the canonical model of X_0, that is, the surface obtained by collapsing all (-2)-curves in X_0 (see Tsuji [**Ts**]).

When algebro-geometers talk about the "degeneration", they also consider the cases that are not covered by Theorem 1.3; namely, the diameter of (X_i, g_i) may diverge. (For example, in the above case of K3 surfaces, consider a sequence that corresponds to a divergent sequence in $K\Omega$.) We do not understand what happens differential geometrically when the diameter goes to infinity. (But see the work of R. Kobayashi [**Kor1**].) If we consider the moduli space of constant curvature metrics on Riemann surfaces, the limit space is always smooth under the assumption on the diameter. The stable curve does not appear! On the other hand, when $k = 1$, i.e., when we consider Einstein-Kähler metrics on del Pezzo surfaces,[8] the diameter condition follows from Myers' theorem as in the remark above, so all the conditions in Theorem 1.3 are satisfied. Thus we get the following:

COROLLARY 1.6. *Fix an underlying differentiable structure $\mathscr{X}$ of the del Pezzo surface X. Let $\mathfrak{M}(\mathscr{X})$ be the moduli space of pairs of complex structures with $c_1 > 0$ and Einstein-Kähler metrics on them. Then adding orbifolds, we can compactify $\mathfrak{M}(\mathscr{X})$.*

Considering only the complex structures, we have a natural map Φ from $\mathfrak{M}(\mathscr{X})$ to the moduli space of complex structures $\mathfrak{H}(\mathscr{X})$. By Bando-Mabuchi's uniqueness theorem for the Einstein-Kähler metric [**BaMa**], the map Φ is injective. In fact, Tian proved the following:

THEOREM 1.7 ([**Ti**]). *If the automorphism group of the del Pezzo surface X is reductive,[9] X has an Einstein-Kähler metric. In other words, the map Φ is bijective unless $\mathscr{X}$ is either the one-point or the two-point blowing-up of $\mathbb{C}P^2$.*

---

[7] An Einstein-Kähler metric means an Einstein metric that is Kählerian at the same time. Then the first Chern class c_1 of the manifold is given by $k[\omega]$ ($[\omega]$ denotes the Kähler class). In particular, if a complex manifold has an Einstein-Kähler metric, either $c_1 > 0$, $= 0$, or < 0.

[8] A complex surface X with $c_1(X) > 0$ is called a del Pezzo surface. It is known that such an X is either $\mathbb{C}P^2$, $\mathbb{C}P^1 \times \mathbb{C}P^1$, or a blowing-up of $\mathbb{C}P^2$ at r generic points ($1 \leq r \leq 8$).

[9] Matsushima's theorem says if a del Pezzo surface admits an Einstein-Kähler metric, its automorphism group is reductive.

The proof is given by showing that the image of Φ is nonempty, both open and closed in $\mathfrak{H}(\mathscr{X})$. It is not so difficult to show the openness. The closedness is shown by using Theorem 1.3. Suppose that (X_i, g_i) is a sequence in $\mathfrak{M}(\mathscr{X})$ and $\Phi(X_i, g_i) \in \mathfrak{H}(\mathscr{X})$ converges. We want to show that (X_i, g_i) also converges and it holds
$$\lim_{i\to\infty} \Phi(X_i, g_i) = \Phi(\lim_{i\to\infty}(X_i, g_i)).$$

The difficulty relates to the Hausdorff-ness of the moduli space $\mathfrak{H}(\mathscr{X})$. If a jumping of the complex structure occurs, one cannot expect the above equality. Moreover, the following problem is still open.

PROBLEM 1.8. When does a del Pezzo surface with quotient singularities have an Einstein-Kähler orbifold metric and lie in the boundary of the compactification of the moduli space $\mathfrak{M}(\mathscr{X})$ given in Theorem 1.3 ?

The problem should relate to the Chow stability, but things are not yet clear.[10]

§2. Bubbling out of ALE spaces

In order to study convergence in more detail, we "blow up" the metrics around the singular points by rescaling. This is a standard technique widely used in problems of weak convergence. As before, we first study the case of harmonic maps.

Let f_j be a sequence of harmonic maps as in Theorem 1.1, and suppose that f_j does not converge at x, i.e., $x \in S$. Then $|df_j(x)|$ goes to infinity as $j \to \infty$. Take a normal coordinate system (x_1, x_2) defined on a neighbourhood U of x. Let x_j be a point at which $|df_j|$ attains the maximum in U. We may assume that $\{x_j\}$ converges to x. Define a new coordinate system by $(y_1, y_2) = (|df_j(x_j)|x_1, |df_j(x_j)|x_2)$ and change the metric as $|df_j(x_j)|^2 g$. Note that the maximum of the energy density of f_j in U is normalized to be 1 with respect to the new metric. The neighbourhood $\{(x_1, x_2) \mid |x_1|^2 + |x_2|^2 < \delta\}$ of x is written as $\{(y_1, y_2) \mid |y_1|^2 + |y_2|^2 < |df_j(x_j)|^2 \delta\}$, in the new coordinates, and converges to $\{(y_1, y_2) \in \mathbb{R}^2\}$ as $j \to \infty$. The rescaled metrics converge to the standard metric on $\mathbb{R}^2$. We consider the map f_j defined on the y-plane and denote it by $\widetilde{f}_j$. Then the energy of the $\widetilde{f}_j$ is bounded from above independently of j. Then one can apply Theorem 1.1 to $\widetilde{f}_j$. In this case, we have a uniform bound on the energy density, so the concentration of the energy does not happen and we get the following:

THEOREM 2.1 ([SaUh]). *There exists a subsequence of $\{\widetilde{f}_j\}$, also denoted by $\{\widetilde{f}_j\}$, which converges to a harmonic map $\widetilde{f}_\infty$ defined on the whole plane $\mathbb{R}^2$ with finite energy.*

Noticing that $\mathbb{R}^2$ and $S^2 \setminus \{p\}$ are conformal to each other, and that energy and harmonicity are preserved under a conformal transformation, we can regard $\widetilde{f}_\infty$ as a finite energy harmonic map defined on $S^2 \setminus \{p\}$. Then by the removable singularities theorem, $\widetilde{f}_\infty$ can be extended to the whole S^2. Also note that $\widetilde{f}_\infty$ depends on $x \in S$.

Let us use the same technique in the case of Einstein metrics. Let $\{(X_j, g_j)\}$ be a sequence of Einstein metrics as in Theorem 1.3, and take $x \in S$. The absolute value of the curvature $|R_{g_j}|$ diverges to infinity at x. Let r_j be the value of $|R_{g_j}|$ at x. Since x is a point in X, not in X_j, we cannot talk about $|R_{g_j}|$ at x. Precisely speaking, we

---

[10] There is progress with this problem since the writing of the original manuscript. See the references added in translation.

mean $|R_{g_j}(x_j)|$, where x_j is a point in X_j and $\{x_j\}$ converges to $x \in X$ in a certain sense (we do not give the precise meaning here to avoid technical complexity). Now we multiply the metric g_j by r_j. Then the absolute value of the curvature of the new metric $r_j g_j$ is normalized to be 1. This rescaling procedure means "viewing by the microscope". Since the curvature becomes uniformly bounded, we can expect convergence of the manifolds. But the diameter goes to infinity, so the limit space becomes a noncompact manifold. In fact, we have the following:

THEOREM 2.2 ([**Na1**]). *Consider a sequence* $\{(X_j, r_j g_j, x_j)\}$ *of the pairs of Riemannian manifolds and points on it. There exists a subsequence, also denoted by* $\{(X_j, r_j g_j, x_j)\}$, *which converges*[11] *to a complete Ricci-flat manifold* (M, h, o) *having the following properties*:

(1) $\int_M |R_h|^2 dV_h < \infty$,

(2) $\operatorname{vol} B(o; r) \geq V r^4$ *for some* $o \in M$, $V > 0$,

where $B(o; r)$ *denotes the ball of radius* r *centered at* o. *More precisely, there exists a diffeomorphism* G_j *from* M *to a neighbourhood of* x_j *such that* $G_j^*(r_j g_j)$ *converges to* h *in* $C^\infty_{\mathrm{loc}}(M)$.

We call these phenomena, such as a map from S^2 being cut off from sequences of harmonic maps, a part of the manifold being torn off in our case, "bubbling out".

In the case of harmonic maps, one can regard the limit map $\widetilde{f}_\infty$ as a map from S^2 through the conformal compactification of $\mathbb{R}^2$. The Einstein equation is not preserved under conformal transformations, so one cannot apply the removable singularities theorem in the same way. However, one can modify its proof to this situation to show the following (in fact, the proof becomes more difficult).

THEOREM 2.3 ([**BaKaNa**]). *If a 4-dimensional complete Ricci-flat manifold satisfies the conditions* (1) *and* (2) *in Theorem* 2.2, *then it is ALE of order four.*

We say a 4-dimensional Riemannian manifold is ALE of order τ if there exists a compact set $K \subset X$ and a finite subgroup $\Gamma \subset \mathrm{SO}(4)$ and a diffeomorphism (coordinates at infinity) $\mathfrak{X}: X \setminus K \to (\mathbb{R}^4 \setminus \overline{B_R})/\Gamma$ such that the following holds in the coordinates $\mathfrak{X}$:

$$|\overbrace{\partial \cdots \partial}^{p \text{ times}}(g_{ij}(x) - \delta_{ij})| = O(|x|^{-p-\tau}) \quad \text{for } x \in (\mathbb{R}^4 \setminus \overline{B_R})/\Gamma, \ p = 0, 1, 2, 3, \ldots.$$

In the orbifold, the distance sphere around a singular point of sufficiently small radius is diffeomorphic to S^3/Γ, and the distance sphere of sufficiently large radius is diffeomorphic to S^3/Γ in the ALE space. Thus one can understand that the ALE space is the counterpart to the orbifold in the category of noncompact manifolds.

REMARK 2.4. One cannot drop condition (2) in Theorem 2.3. There exists a Ricci-flat Kähler metric, called the Taub-NUT metric, defined on $\mathbb{C}^2$. This metric has an asymptotic behaviour called ALF, and the volume of a ball of radius r grows as $O(r^3)$ (see [**EGH**]). We do not have examples of Ricci-flat manifolds satisfying condition (2) but not satisfying (1).

We shall see what happens in Example 1.5. Take a point y_n in an exceptional set E_n in X that corresponds to a singular point x_n in X_∞. If the value of $|R_{g_i}|$ at

---

[11] with respect to the pointed Hausdorff convergence

y_n is equal to r_i, $(X, r_i g_i)$ converges to an ALE Ricci-flat Kähler manifold (M, h). This M is called the Eguchi-Hanson space and is biholomorphic to the cotangent bundle $T^*\mathbb{CP}^1$ of the complex projective line. ($T^*\mathbb{CP}^1$ is the minimal resolution of the simple singularity of type A_1. The exceptional set is the zero section.) In this case the limit space is independent of n. In general, when K3 surfaces degenerate to an orbifold with simple singularities, one can obtain a Ricci-flat Kähler metric on the minimal resolution of each singularity in the same way (see [**Kor2**]).

In Theorems 2.2 and 2.3, we rescale the metric so that the curvature is uniformly bounded. We lose some information and need more detailed study in general. This is because several ALE spaces may blow up from one singular point $x \in S \subset X_\infty$. An accurate version of the convergence theorem which includes the above situations was obtained by Bando [**Ba**].[12] In order to explain his result, we need some terminology and symbols. We say an orbifold metric is ALE when there is no singular set outside a compact set, and there exists a coordinate system $\mathfrak{X}$ at infinity as usual in ALE Riemannian manifolds. For such an ALE orbifold (M, h), we denote by $\mathcal{S}(M)$ the set of singular points. By definition, $\mathcal{S}(M)$ is a finite set. We denote by x_j^n the point in X_j that converges to the singular point x^n in X_∞ in the same sense as before. (See the explanation above Theorem 2.2.)

We see a neighbourhood of x_j^n through a microscope. In this case, we set the magnification r_j so that

$$\int_{B(x_j^n; R) \setminus B(x_j^n; \frac{1}{\sqrt{r_j}})} |R_{g_j}|^2 dV_{g_j} = \varepsilon,$$

where ε and R are (sufficiently small) positive constants. Here $B(x; r)$ denotes the metric ball of radius r centered at x. The convergence theorem similar to Theorem 1.3 can be applied to the magnified Riemannian manifold $(X_j, h_j) = (X_j, r_j g_j)$. Since we have

$$\int_{B(x_j^n; \sqrt{r_j} R) \setminus B(x_j^n; 1)} |R_{h_j}|^2 dV_{h_j} = \varepsilon$$

in this case, the concentration of the curvature does not happen in $B(x_j^n; \sqrt{r_j} R) \setminus B(x_j^n; 1)$; hence the limit (M, h) of (X_j, h_j) is an ALE orbifold that does not have singularities outside the ball of radius 1. If we want to know the situation of the convergence, we apply the same argument to (X_j, h_j) instead of (X_j, g_j). Namely, we multiply h_j again around a singular point of (M, h). This means the magnification is reset bigger. We make the magnification bigger as a singular point occurs in the limit ALE Ricci-flat orbifold in this way; then finally, no singular points appear and the procedure ends in a finite number of steps. We adjust the following form. The statement becomes complicated, so we give Figure 2.1 on p. 88 to describe the circumstances.

THEOREM 2.5 ([**Ba**]). (1) *In the same notation as in Theorem 1.3* (*taking a further subsequence* $\{j\} \subset \{i\}$ *if necessary*), *there exist a family of pointed ALE Ricci-flat orbifolds and sequences of positive numbers that have a recursive relation as follows:*

---

[12] This paper was dedicated to the late cartoonist Osamu Tezuka.

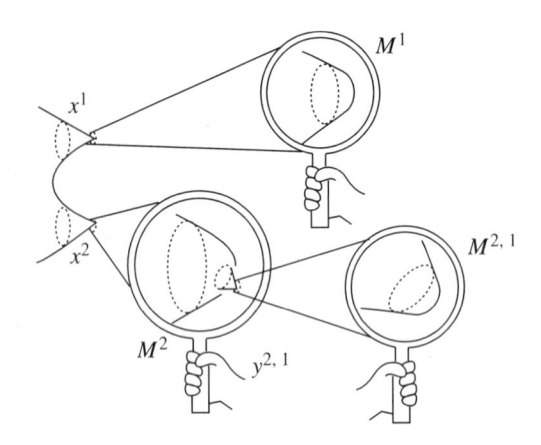

[FIGURE 2.1]

$\boxed{1}$ *There exist a pair* (M^n, h^n, o^n) *of an ALE Ricci-flat orbifold and a point corresponding to each singular point* $x^n \in S$ *in* X_∞ ($n = 1, 2, \ldots, \#S$) *and a divergent sequence* $\{r_j^n\}_{j=1,2,\ldots}$ *of positive numbers such that*

$$\lim_{j\to\infty} (X_j, r_j^n g_j, x_j^n) = (M^n, h^n, o^n).$$

Here lim *is convergence in the sense of Theorem 2.2. Moreover, the fundamental group of the end of* M^n *and the group corresponding to the singular point* x^n *are isomorphic and their actions on* $\mathbb{R}^4$ *are the same.*

$\boxed{k-1 \Rightarrow k}$ *Suppose that the ALE Ricci-flat orbifold* $M^{n_1,n_2,\ldots,n_{k-1}}$ *is defined and has a singular point* $y^{n_1,n_2,\ldots,n_k} \in \mathcal{S}(M^{n_1,n_2,\ldots,n_{k-1}})$ ($n_k = 1, 2, \ldots, \#\mathcal{S}(M^{n_1,n_2,\ldots,n_{k-1}})$). *There exist a pair* $(M^{n_1,n_2,\ldots,n_k}, h^{n_1,n_2,\ldots,n_k}, o^{n_1,n_2,\ldots,n_k})$ *consisting of an ALE Ricci-flat orbifold and a point, a sequence* $\{r_j^{n_1,n_2,\ldots,n_k}\}_{j=1,2,\ldots}$ *of positive numbers, and a sequence* $\{x_j^{n_1,n_2,\ldots,n_k}\}_{j=1,2,\ldots}$ *of points in* X_j *such that*

$$\lim_{j\to\infty} (X_j, r_j^{n_1,n_2,\ldots,n_k} g_j, x_j^{n_1,n_2,\ldots,n_k}) = (M^{n_1,n_2,\ldots,n_k}, h^{n_1,n_2,\ldots,n_k}, o^{n_1,n_2,\ldots,n_k}),$$

$$\lim_{j\to\infty} \frac{r_j^{n_1,n_2,\ldots,n_k}}{r_j^{n_1,n_2,\ldots,n_{k-1}}} = \infty.$$

Moreover, the fundamental group of the end of $M^{n_1,n_2,\ldots,n_k}$ *and the group corresponding to the singular point* $y^{n_1,n_2,\ldots,n_k}$ *are isomorphic and their actions on* $\mathbb{R}^4$ *are the same.*

(2) *This recursive procedure ends in a finite number of steps; that is, the ALE orbifold* $M^{n_1,n_2,\ldots,n_k}$ *becomes nonsingular for some* k. *Moreover, for sufficiently large* j *and sufficiently small* R, *the ball* $B(X_j^n; R)$ *is diffeomorphic to a manifold which is obtained from* $M^{n_1,n_2,\ldots,n_k} \setminus \mathcal{S}(M^{n_1,n_2,\ldots,n_k})$*'s with* $n_1 = n$ *by attaching a neighbourhood of* $y^{n_1,n_2,\ldots,n_k}$ *and the end of* $M^{n_1,n_2,\ldots,n_k}$.

(3) *The measure* $|R_{g_j}|^2 dV_{g_j}$ *converges to*

$$|R_{g_\infty}|^2 dV_{g_\infty} + \sum_{n=1}^{\#S} a_n \delta_{x^n},$$

where a_n *is given by*

$$a_n = \sum_{(M,h)} \int_M |R_h|^2 dV_h.$$

Here the summation runs over the set of ALE orbifolds $(M, h) = (M^{n_1, n_2, \ldots, n_k}, h^{n_1, n_2, \ldots, n_k})$ *with* $n_1 = n$.

For a compact Einstein 4-manifold, the L^2-norm of the curvature gives its Euler number. For an ALE Einstein manifold (M, h) we have

$$\frac{1}{8\pi^2} \int_M |R_h|^2 dV_h = \text{the Euler number of } M - \frac{1}{\text{the order of } \Gamma}.$$

(There are similar formulae for compact orbifolds and ALE orbifolds.) We have the following. (For the definition of hyper-Kähler structure, see §3.)

PROPOSITION 2.6. *Under the same situation as in Theorem 1.3, suppose that* (X_i, g_i) *has a Kähler (resp. hyper-Kähler) structure. Then the limit space* (X_∞, g_∞) *and the space* $(M^{n_1, n_2, \ldots, n_k}, h^{n_1, n_2, \ldots, n_k})$ *appearing in Theorem 2.5 also have a Kähler (resp. hyper-Kähler) structure in the sense of orbifolds.*

§3. ALE spaces with hyper-Kähler structures

As we said in the previous section, the ALE spaces bubble out when Einstein metrics degenerate to an orbifold metric. In this way, one obtains existence of ALE spaces. In this section, we shall give a more concrete way of obtaining ALE spaces with hyper-Kähler structures. This is a result by Kronheimer [Kr1, 2]. His result is closely related to the theory of *simple singularities*. A simple singularity is a quotient space $\mathbb{C}^2/\Gamma$, where Γ is a finite subgroup of SU(2), and has been studied for a long time. The classification is known, and it can be realized as a complex hypersurface in $\mathbb{C}^3$. The dual graph of the exceptional set of its minimal resolution is a Dynkin graph with no double edges (i.e., of type A, D, E). (See Table 3.1.) By the theory of Brieskorn [Br] and Slodowy [Sl], one can construct the simple singularity using nilpotent varieties of the corresponding simple Lie algebra. One can also describe the semi-universal deformation and its simultaneous resolution in terms of the Lie algebra. Kronheimer succeeded in constructing these varieties using so-called *quivers*. Moreover, his construction also showed that each fiber has a hyper-Kähler structure.

A *hyper-Kähler structure* on a Riemannian manifold (M, g) is, by definition, a triple of almost complex structures (I, J, K) that satisfies a quaternion relation $IJ = -JI = K$ and is parallel with respect to the Levi-Civita connection of g,

Table 3.1

root system	group	Dynkin graph	hypersurface
A_n	the cyclic group of order $n+1$	o—o—o—⋯—o	$x^{n+1} + yz = 0$
D_n	the binary dihedral group of order $4(n-2)$	o—o—⋯—o<	$x^2 + y^2z + z^{n-1} = 0$
E_6	the binary tetrahedral group	o—o—o—o—o	$x^2 + y^3 + z^4 = 0$
E_7	the binary octahedral group	o—o—o—o—o—o	$x^2 + y^3 + yz^3 = 0$
E_8	the binary icosahedral group	o—o—o—o—o—o—o	$x^2 + y^3 + z^5 = 0$

i.e., $\nabla I = \nabla J = \nabla K = 0$. We call a Riemannian manifold with a hyper-Kähler structure simply a hyper-Kähler manifold. The holonomy group of a hyper-Kähler manifold is contained in $\mathrm{Sp}(n)$, where $n = \frac{1}{4} \dim M$. Each almost-complex structure I, J, or K defines a Kähler structure on (M, g). When the manifold is 4-dimensional and simply connected, a hyper-Kähler structure is nothing but a Ricci-flat Kähler structure thanks to the isomorphism $\mathrm{Sp}(1) = \mathrm{SU}(2)$. For example, a Ricci-flat Kähler metric on a K3 surface gives us a hyper-Kähler structure. Taking a particular almost-complex structure I and regarding (M, g) as a Kähler manifold, we have a nowhere vanishing closed $(2, 0)$-form $\omega_J + i\omega_K$, where ω_J (resp. ω_K) is the Kähler form associated with J (resp. K). Namely, we have a *holomorphic symplectic form*. Conversely, if a compact Kähler manifold has a holomorphic symplectic form, there exists a Ricci-flat Kähler metric by the solution of the Calabi conjecture, and hence there exists a hyper-Kähler structure.

On the other hand, Hitchin et al. [**HKLR**] introduced a notion of hyper-Kähler quotients which is an analogue of Marsden-Weinstein quotients for symplectic manifolds, therefore giving a different method for constructing hyper-Kähler manifolds. Let us recall their result.

Suppose that a Lie group G acts on a hyper-Kähler manifold (M, g, I, J, K) preserving the hyper-Kähler structure. Let us denote the Lie algebra of G by $\mathfrak{g}$, and its dual space by $\mathfrak{g}^*$. G acts on $\mathfrak{g}^*$ by the co-adjoint action. A G-equivariant map $\mu = (\mu_I, \mu_J, \mu_K) \colon M \to \mathbb{R}^3 \otimes \mathfrak{g}^*$ is said to be a *hyper-Kähler moment map* if we have

$$I \operatorname{grad}\langle \mu_I, \xi \rangle = J \operatorname{grad}\langle \mu_J, \xi \rangle = K \operatorname{grad}\langle \mu_K, \xi \rangle = \xi^*$$

for any $\xi \in \mathfrak{g}$. Here $\langle\ ,\ \rangle$ is a natural pairing between $\mathfrak{g}$ and $\mathfrak{g}^*$, and ξ^* is a vector field generated by ξ. Let us define

$$Z = \{\zeta \in \mathfrak{g}^* \mid \operatorname{Ad}_g^*(\zeta) = \zeta \text{ for any } g \in G\}.$$

Then $\mu^{-1}(\zeta)$ is invariant under the G-action for $\zeta \in \mathbb{R}^3 \otimes Z$. So we can consider the quotient space $X_\zeta = \mu^{-1}(\zeta)/G$.

THEOREM 3.1 ([**HKLR**]). *Suppose that the G-action on $\mu^{-1}(\zeta)$ is free. Then the quotient space X_ζ is a C^∞-manifold and has a Riemannian metric and a hyper-Kähler structure induced from those on M.*

As an application of the above hyper-Kähler quotient construction, we give ALE spaces. Fix a root system of type A, D, or E. Let $\theta_1, \ldots, \theta_r$ be its simple roots. Let $\theta_0 = -\sum_i n_i \theta_i$ be the negative of the highest root. Set $n_0 = 1$ for convenience. Draw the extended Dynkin diagram, and put a complex vector space of dimension n_i on each vertex i. Consider linear maps $f_{ij} \colon \mathbb{C}^{n_i} \to \mathbb{C}^{n_j}$ and $f_{ji} \colon \mathbb{C}^{n_j} \to \mathbb{C}^{n_i}$ when the vertices i and j are joined by an edge. See Figure 3.1 for D_5.

Let M be the linear space of all such linear maps f_{ij}. Put a Hermitian metric on each $\mathbb{C}^{n_i}$ and also the induced metric on M. To make M a quaternion vector space, we introduce J as follows. For joined vertices i and j, we have two linear maps from $\mathbb{C}^{n_i}$ to $\mathbb{C}^{n_j}$ and $\mathbb{C}^{n_j}$ to $\mathbb{C}^{n_i}$. Choose one of them and fix henceforth. When $i \to j$ is chosen, we define J by

$$J(f_{ij}, f_{ji}) = (-f_{ji}^*, f_{ij}^*),$$

where $*$ means the adjoint. Then J satisfies $J^2 = -1$ and is antilinear with respect to the complex structure. Hence M has a structure of a quaternion vector space. There

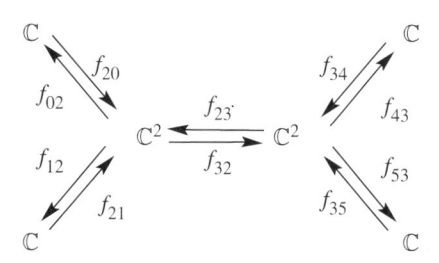

FIGURE 3.1

is a natural action on M of a Lie group $G = \prod_{i \neq 0} \mathrm{U}(\mathbb{C}^{n_i})$.[13] This action preserves the metric and the quaternion structure. Let μ be the hyper-Kähler moment map vanishing at the origin, which exists uniquely. Let Z be as above, and take $\zeta \in \mathbb{R}^3 \otimes Z$. Then the space $X_\zeta = \mu^{-1}(\zeta)/G$ has a hyper-Kähler structure on its nonsingular part. More precisely, Kronheimer showed the following:

THEOREM 3.2 ([Kr1]). *For a generic ζ, the G-action on $\mu^{-1}(\zeta)$ is free and the quotient space X_ζ is a nonsingular 4-dimensional hyper-Kähler manifold. Moreover, X_ζ is diffeomorphic to a minimal resolution of $\mathbb{C}^2/\Gamma$ and the metric is ALE.*

He had a precise description of X_ζ even for nongeneric ζ; for example, $X_\zeta = \mathbb{C}^2/\Gamma$ if $\zeta = 0$. Write $\zeta = (\zeta_I, \zeta_J, \zeta_K)$, and consider the following family of spaces with $\zeta_I = 0$,

$$\bigcup_{\zeta_J + i\zeta_K \in \mathbb{C} \otimes Z} X_{(0, \zeta_J, \zeta_K)},$$

and the natural projection to $\mathbb{C} \otimes Z$. Then it is a semiuniversal deformation of $\mathbb{C}^2/\Gamma$ pulled back to the Weyl group covering. For generic ζ_I, one can define a holomorphic map

$$X_{(\zeta_I, \zeta_J, \zeta_K)} \to X_{(0, \zeta_J, \zeta_K)},$$

which is a simultaneous resolution of singularities. When a sequence of generic parameters converges to a nongeneric parameter, corresponding Einstein metrics converge to an ALE orbifold as in §§1,2.

The geometric meaning of the parameter ζ can be explained in terms of period maps. The exceptional set of the resolution of $\mathbb{C}^2/\Gamma$ is a union of $\mathbb{C}P^1$'s and their intersection matrix is -1 times the Cartan matrix of the corresponding Lie algebra. Then for generic ζ, one can identify $H^2(X_\zeta; \mathbb{R})$ with the Cartan subalgebra $\mathfrak{h}$, and an element Σ in $H_2(X_\zeta; \mathbb{Z})$ with $\Sigma \cdot \Sigma = -2$ corresponds to a root. The set $(\mathbb{R}^3 \otimes Z)^\circ$ of generic ζ is simply connected; so the local system $H^2(X_\zeta; \mathbb{R})$ can be trivialized. So taking cohomology classes of three Kähler forms ω_I, ω_J, ω_K on X_ζ, we can define a period map

$$P: (\mathbb{R}^3 \otimes Z)^\circ \to \mathbb{R}^3 \otimes \mathfrak{h}.$$

Then Kronheimer showed that P is induced from an isomorphism between Z and

---

[13] The action of $\prod \mathrm{U}(\mathbb{C}^{n_i})$ is not appropriate, since $c\,\mathrm{Id} \in \prod \mathrm{U}(\mathbb{C}^{n_i})$ fixes any element in M.

$\mathfrak{h}$, and $\zeta \in \mathbb{R}^3 \otimes Z$ is generic (i.e., X_ζ is smooth) if and only if

$$P(\zeta) \in \bigcup_\theta \mathbb{R}^3 \otimes \pi_\theta,$$

where π_θ is a hypersurface defined by the root θ.

He also showed the following classification theorem:

THEOREM 3.3 ([**Kr2**]). *An ALE space with a hyper-Kähler structure is isomorphic to the one given by the above construction.*

As above, an ALE space is well understood when it has a hyper-Kähler structure. But it also seems important to study the spaces with Kähler structures in conjunction with algebraic geometry. In this case an ALE space is a cyclic quotient of an ALE space with a hyper-Kähler structure [**Ba**]. Using the above classification theorem and an argument in [**Ti**], we can show the following:

THEOREM 3.4. *Let (X, g) be a nonsingular ALE Ricci-flat Kähler 4-manifold. Then one of the following holds.*

(1) *(X, g) has a hyper-Kähler structure compatible with the Kähler structure.*

(2) *(X, g) is a quotient of a hyper-Kählerian ALE space $\widetilde{X}$ of type A_n by the cyclic group $\mathbb{Z}/r\mathbb{Z}$, and $n + 1$ is a multiple of r.*

As we saw in §§1,2, an ALE space with a Kähler structure bubbles out when Einstein-Kähler metrics converge to an orbifold. For example, consider a deformation of an algebraic surface $\pi: \mathfrak{X} \to \Delta$ and suppose that $X_t = \pi^{-1}(t)$ has an Einstein-Kähler metric g_t when $t \neq 0$. If the diameter of (X_t, g_t) is estimated from the above independently of t, the convergence theorem gives us an Einstein-Kähler orbifold as a limit. By the above theorem together with the relation between the singularity and the bubbling out of ALE spaces, the singularity of the orbifold must be a simple singularity or a cyclic quotient of a simple singularity of type A_n. Strikingly, the similar result is obtained by using the minimal model theory (see Kawamata [**Ka**]). If the total space $\mathfrak{X}$ has only terminal singularities and the central fiber X_0 has only orbifold singularities, then the singularities in the central fiber are of the type just mentioned.

At the end, we give the following problem:

PROBLEM 3.5. *Are there ALE Ricci-flat 4-manifolds without Kähler structures?*

We do not have such examples at this moment.

References

[An1]   M. Anderson, *Ricci curvature bounds and Einstein metrics on compact manifolds*, J. Amer. Math. Soc. **2** (1989), 455–490.

[An2]   _____, *The L^2 structure of moduli spaces of Einstein metrics on 4-manifolds*, Geom. Funct. Anal. **2** (1992), 29–89.

[Au]   T. Aubin, *Équations du type de Monge-Ampère sur les variétés kählériennes compactes*, C. R. Acad. Sci. Paris **283** (1976), 119–121.

[BaCo]   A. Bahri and J. Coron, *On a nonlinear elliptic equation involving the critical exponent: The effect of the topology of the domain*, Comm. Pure Appl. Math. **41** (1988), 253–294.

[Ba]   S. Bando, *Bubbling out of Einstein manifolds*, Tôhoku Math. J. (2) **42** (1990), 205–216.

[BaKaNa]   S. Bando, A. Kasue, and H. Nakajima, *On a construction of coordinates at infinity on manifolds with fast curvature decay and maximal volume growth*, Invent. Math. **97** (1989), 313–349.

[BaMa] S. Bando and T. Mabuchi, *Uniqueness of Einstein-Kähler metrics modulo connected group actions*, Algebraic Geometry, Sendai, 1985, Adv. Stud. Pure Math., vol. 10, Kinokuniya, Tokyo, 1987, pp. 11–40.
[Be] A. Besse, *Einstein manifolds*, A Series of Modern Surveys in Mathematics, Band 10, Springer-Verlag, Berlin and Heidelberg, 1987.
[BrCo] H. Brezis and J. Coron, *Convergence of solutions of H systems or how to blow bubbles*, Arch. Rat. Mech. Anal. **89** (1985), 21–56.
[Br] E. Brieskorn, *Singular elements of semisimple algebraic groups*, Proc. Internat. Congr. Math. (Nice, 1970), Vol. 2, Gauthier-Villars, Paris, 1970, pp. 279–284.
[Do1] S. K. Donaldson, *Anti-self-dual Yang-Mills connections over complex algebraic surfaces and stable vector bundles*, Proc. London Math. Soc. **50** (1985), 1–26.
[Do2] _____, *Infinite determinants, stable bundles and curvature*, Duke Math. J. **54** (1987), 231–247.
[EgGiHa] T. Eguchi, P. Gilkey, and A. Hanson, *Gravitation, gauge theories and differential geometry*, Phys. Rev. **66** (1980), 215–393.
[EgHa] T. Eguchi and A. Hanson, *Asymptotically flat solutions to Euclidean gravity*, Phys. Lett. **B 74** (1978), 249–251.
[Ga] L. Gao, *Einstein metrics*, J. Diff. Geom. **32** (1990), 155–183.
[GiHa] G. Gibbons and S. Hawking, *Gravitational multi-instantons*, Phys. Lett. **B 78** (1978), 430–432.
[Hi] N. J. Hitchin, *Polygons and gravitons*, Math. Proc. Cambridge Philos. Soc. **85** (1979), 465–476.
[HKLR] N. J. Hitchin, A. Karhede, U. Lindström, and M. Roček, *Hyperkähler metrics and supersymmetry*, Comm. Math. Phys. **108** (1987), 535–589.
[Ka] Y. Kawamata, *Crepant blowing-up of 3-dimensional canonical singularities and its applications to degenerations of surfaces*, Ann. of Math. (2) **127** (1988), 93–163.
[Kor1] R. Kobayashi, *Ricci-flat Kähler metrics on affine algebraic manifolds and degenerations of Kähler-Einstein K3 surfaces*, Kähler Metrics and Moduli Spaces, Adv. Stud. Pure Math., vol. 18-II, Kinokuniya, Tokyo, 1990, pp. 137–228.
[Kor2] _____, *Moduli of Einstein metrics on K3 surfaces and degeneration of type* I, Kähler Metrics and Moduli Spaces, Adv. Stud. Pure Math., vol. 18-II, Kinokuniya, Tokyo, 1990, pp. 257–311.
[KorTo] R. Kobayashi and A. Todorov, *Polarized period map for generalized K3 surfaces and the moduli of Einstein metrics*, Tôhoku Math J. (2) **39** (1987), 145–151.
[Kos] S. Kobayashi, *Curvature and stability of vector bundles*, Proc. Japan Acad. **58** (1982), 158–162.
[Kr1] P. B. Kronheimer, *The construction of ALE spaces as hyper-Kähler quotients*, J. Differential Geom. **29** (1989), 665–683.
[Kr2] _____, *A Torelli-type theorem for gravitational instantons*, J. Differential Geom. **29** (1989), 685–697.
[KrNa] P. B. Kronheimer and H. Nakajima, *Yang-Mills instantons on ALE gravitational instantons*, Math. Ann. **288** (1990), 263–307.
[Lü] M. Lübke, *Stability of Einstein-Hermitian vector bundles*, Manuscripta Math. **42** (1983), 245–257.
[Mu] S. Mukai, *On the moduli space of bundles on K3 surfaces*. I, Vector Bundles on Algebraic Varieties, Oxford Univ. Press, 1987, pp. 341–413.
[Na1] H. Nakajima, *Hausdorff convergence of Einstein 4-manifolds*, J. Fac. Sci. Univ. Tokyo **35** (1988), 411–424.
[Na2] _____, *Moduli spaces of anti-self-dual connections on ALE gravitational instantons*, Invent. Math. **102** (1990), 267–303.
[NaSe] M. S. Narashimhan and C. S. Seshadri, *Stable and unitary vector bundles on compact Riemann surfaces*, Ann. of Math. (2) **82** (1965), 540–567.
[Pa] D. Page, *A physical picture of the K3 gravitational instantons*, Phys. Lett. **B 80** (1978), 55–57.
[SaUh] J. Sacks and K. Uhlenbeck, *The existence of minimal immersion of 2-spheres*, Ann. of Math. (2) **113** (1981), 1–24.
[Se] S. Sedlacek, *A direct method for minimizing the Yang-Mills functional*, Comm. Math. Phys. **86** (1982), 515–528.
[Si] Y. T. Siu, *Lectures on Hermitian-Einstein metrics for stable bundles and Kähler-Einstein metrics*, DMV Seminar, Band 8, Birkhäuser-Verlag, Basel and Boston, 1987.
[Sl] P. Slodowy, *Simple singularities and simple algebraic groups*, Lecture Notes in Math., vol. 815, Springer-Verlag, Berlin, Heidelberg and New York, 1980.
[Ti] G. Tian, *On Calabi's conjecture for complex surfaces with positive first Chern class*, Invent. Math. **101** (1990), 101–172.

[Ts]   G. Tsuji, *Existence and degeneration of Kähler-Einstein metrics on minimal algebraic varieties of general type*, Math. Ann. **281** (1988), 123–133.
[Uh1] K. Uhlenbeck, *Removable singularities in Yang-Mills fields*, Comm. Math. Phys. **83** (1982), 11–30.
[Uh2] _____, *Connection with L^p-bounds on curvature*, Comm. Math. Phys. **83** (1982), 31–42.
[UhYa] K. Uhlenbeck and S. T. Yau, *On the existence of Hermitian-Yang-Mills connections in stable vector bundles*, Comm. Pure Appl. Math. **39(S)** (1986), 258–293.
[Ya]   S. T. Yau, *On the Ricci curvature of a compact Kähler manifold and the complex Monge-Ampère equation*. I, Comm. Pure Appl. Math. **31** (1978), 339–411.

Added in translation

[DiTi] W. Ding and G. Tian, *Kähler-Einstein metrics and the generalized Futaki invariant*, Invent. Math. **110** (1992), 315–335.
[MaMu] T. Mabuchi and S. Mukai, *Stability and Einstein-Kähler metric of a quartic del Pezzo surface*, in Einstein Metrics and Yang-Mills Connections (T. Mabuchi and S. Makai, eds.), Lecture Notes Pure Appl. Math., vol. 145, Marcel Dekker, New York, 1993, pp. 133–160.

MATHEMATICAL INSTITUTE, TÔHOKU UNIVERSITY, ARAMAKI, AOBA-KU, SENDAI 980, JAPAN
E-mail address: nakajima@math.tohoku.ac.jp

Translated by HIRAKU NAKAJIMA

On the Topology of Elliptic Surfaces—a Survey

Masaaki Ue

A complex surface X is called an *elliptic surface* if it admits a complex analytic projection $\pi: X \to B$ onto some Riemann surface B such that its general fiber is a nonsingular elliptic curve (a 2-torus T^2). Here the inverse image by π of a critical value of π is called a *singular fiber* and the inverse image by π of another point is called a *general fiber*. Elliptic surfaces have structures that can be described explicitly by the study of Kodaira [K1, K2], etc. and also have some interesting properties as 4-manifolds. For example, applications of the theories of Freedman and Donaldson show that some classes of elliptic surfaces provide us infinitely many distinct smooth structures on the same closed topological manifolds. Such phenomena cannot occur in the other dimensions. Furthermore, we can expect that elliptic surfaces give some hints for producing interesting 4-manifolds whose properties can be clarified only by the above theories. The purpose of this note is to fill in the gaps between the basic properties of elliptic surfaces and the deeper results about them. We will describe the fiber structures on elliptic surfaces in §1, their fundamental properties as complex surfaces in §2, their homeomorphism and diffeomorphism types in §§4–6, and the decompositions of elliptic surfaces and some related topics (including Gompf-Mrowka's "homotopy $K3$s" that are not complex surfaces) in §7. Section 3 is devoted to giving the results about Seifert fibered 3-manifolds which will be needed in the later sections. In this note we will identify all mutually diffeomorphic manifolds. The complex structures on the complex surfaces themselves are not taken into account, although we fix the orientations derived from them.[1] So we will denote any 4-manifold diffeomorphic to an elliptic surface simply by an elliptic surface. All the manifolds will be smooth, oriented, and compact unless otherwise specified.

NOTATION.

S^n : n-sphere, $\quad B^n$: n-ball, $\quad T^n$: n-torus,

$\mathbf{CP}^n$: the complex projective space of dimension n.

1991 *Mathematics Subject Classification.* Primary 14J27.

This article originally appeared in Japanese in Sūgaku **44** (3) (1992), 205–228.

[1] See the book [FM4] for a comprehensive study of elliptic surfaces including their deformation types.

For oriented 4-manifolds X, X'

$aX \cong \overbrace{X \sharp \cdots \sharp X}^{a}$: a connected sum of a copies of X,

$\overline{X}$ or $-X$: X with its orientation reversed, ∂X: the boundary of X,

$q_X : H_2(X, \mathbf{Z})/\text{Torsion} \times H_2(X, \mathbf{Z})/\text{Torsion} \longrightarrow \mathbf{Z}$:

the intersection form of X: $(q_X(x, y) = x \cdot y)$,

$b_i(X)$: the ith Betti number of X,

$b_2^\pm(X)$: the number of positive (negative) eigenvalues of q_X,

$\sigma(X) = b_2^+(X) - b_2^-(X)$, $e(X)$: the Euler number of X,

$X \cong X'$: there is an oriented diffeomorphism between X and X'.

The negative-definite nonsingular bilinear form over $\mathbf{Z}$ of type II, rank 8 is denoted by E_8, and the bilinear form represented by the matrix $\begin{pmatrix} 0 & 1 \\ 1 & 0 \end{pmatrix}$ is denoted by U.

§1. The structures of fibrations for elliptic surfaces

First let us fix the projection $\pi : X \to B$ for an elliptic surface and describe its fibering structure from the topological point of view. Here we assume that X is compact and so the number of singular fibers is finite.

DEFINITION (1-1). The elliptic surface $\pi : X \to B$ is called *relatively minimal* (or simply *minimal*) if any fiber of π does not contain an exceptional curve of the first kind (a nonsingular rational curve C with $C \cdot C = -1$).

If the fibration π contains k exceptional curves of the first kind in its fibers, then by collapsing each such curve into one point (blow-down) we get a minimal elliptic surface X' and a new projection $\pi' : X' \to B$ such that

$$X \cong X' \sharp k \overline{\mathbf{CP}}^2.$$

Hereafter we will only consider minimal elliptic surfaces unless otherwise specified.

DEFINITION (1-2). Take a disk neighborhood Δ of a point p in B so that Δ contains no images of the singular fibers by π other than p. Then $\pi^{-1}(\Delta)$ is a regular neighborhood of the fiber $F = \pi^{-1}(p)$ over p. We will denote it by $N(F)$ and the restriction of π to $N(F)$ simply by $\pi : N(F) \to \Delta$.

Pick up all singular fibers F_i of $\pi : X \to B$, and let B_0 be the resulting manifold obtained from B by removing all small disk neighborhoods Δ_i of $\pi(F_i)$. Then $X_0 = \pi^{-1}(B_0)$ is a T^2-bundle over B_0. Moreover, X is described by the following information.
 (1) The T^2-bundle $\pi' : X_0 \to B_0$ over B_0.
 (2) The neighborhood of some singular fiber F_i for an elliptic surface and the projection $\pi_i : N(F_i) \to \Delta_i$ in the sense of (1-2).

(3) $X = X_0 \cup (\bigcup_{\varphi_i} N(F_i))$, $B = B_0 \cup (\bigcup \Delta_i)$, $\pi|_{X_0} = \pi'$, $\pi|_{N(F_i)} = \pi_i$.
(4) The map φ_i is a fiber-preserving diffeomorphism from $\partial N(F_i)$ to $\pi^{-1}(\partial \Delta_i) \subset X_0$.

DEFINITION (1-3). We call a 4-manifold with a projection $\pi : X \to B$ (B is a 2-manifold) defined by (1)–(4) above a *singular torus fibration*. This definition is also valid for the cases when $\partial B \neq \emptyset$.

In [**Mt1, Mt3**] 4-manifolds with fibrations that admit wider classes of singular fibers are called torus fibrations, but here we only consider the fibrations whose singular fibers appear in some elliptic surface. Obviously, any elliptic surface is a singular torus fibration in the above sense, but we will see in §5 that the converse is not true (some of the singular torus fibrations do not admit complex structures). Take a base point $*$ in B_0 and a simple closed curve γ through $*$. Then $\pi : \pi^{-1}(\gamma) \to \gamma$ is a T^2-bundle over S^1 of the form

$$\pi^{-1}(\gamma) \cong [0,1] \times T^2/(1,x) \sim (0,f(x)) \quad \text{for } x \in T^2.$$

Here f is an orientation-preserving self-diffeomorphism of T^2. As far as the diffeomorphism types are concerned we have only to consider the isotopy class of f, and so we can assume that f is represented by a linear map $A \in \mathrm{SL}(2,\mathbf{Z})$ with respect to some identification of $\pi^{-1}(*) \cong T^2$ with $\mathbf{R}^2/\mathbf{Z}^2$. Then two curves $\ell = (\mathbf{R} \times *)/\mathbf{Z}^2$ and $h = (* \times \mathbf{R})/\mathbf{Z}^2$ (where $*$ is one point) form a basis for $H_1(T^2, \mathbf{Z})$ and A gives a transformation of $H_1(T^2, \mathbf{Z})$. If another basis is chosen then A is replaced by its conjugate.

DEFINITION (1-4). (1) We call (ℓ, h) a *basis* of the general fiber $\pi^{-1}(*)$. (2) The map f defined above is called a *monodromy* of γ and A is called a *monodromy representation* of γ with respect to (ℓ, h). For a singular fiber F and the projection $\pi : N(F) \to \Delta$ defined in (1-2) the monodromy (representation) of $\partial \Delta$ is called the monodromy (representation) of F.

For any fiber F we can take a disk $\widetilde{\Delta}$ that is smoothly embedded in $N(F)$ so that $\widetilde{\Delta}$ intersects every fiber in $N(F)$ transversely, and $F \cap \widetilde{\Delta}$ is one point. Then $\pi : \widetilde{\Delta} \to \Delta$ is a cyclic branched covering branched over one point of Δ.

DEFINITION (1-5). The degree m of $\pi : \widetilde{\Delta} \to \Delta$ is called the *multiplicity* of F, and if $m \geq 2$ we call F a *multiple fiber* of multiplicity m. If $m = 1$, then $\widetilde{\Delta}$ is a cross section of Δ.

On the other hand, the (minimal) singular fibers in elliptic surfaces were classified by Kodaira [**K1, BPV**]. We give their list together with their Euler numbers and monodromy representations with respect to an appropriate basis.

The graphs in Table 1 on p. 98 represent the following divisors:

TABLE 1

Singular fiber		Monodromy	Euler number
$_mI_0$	nonsingular elliptic of multiplicity m	$\begin{pmatrix} 1 & 0 \\ 0 & 1 \end{pmatrix}$	0
$_mI_1$	rational with a node of multiplicity m	$\begin{pmatrix} 1 & 0 \\ 1 & 1 \end{pmatrix}$	1
$_mI_b$ $(\widetilde{A}_{b-1})$	b vertices, each labeled m (cyclic chain)	$\begin{pmatrix} 1 & 0 \\ b & 1 \end{pmatrix}$	b
II	rational with a cusp	$\begin{pmatrix} 1 & -1 \\ 1 & 0 \end{pmatrix}$	2
III $(\widetilde{A}_1)$	two $\mathbf{CP}^1$s C_1, C_2 with $C_1 \cdot C_2 = 2$	$\begin{pmatrix} 0 & -1 \\ 1 & 0 \end{pmatrix}$	3
IV $(\widetilde{A}_2)$	three $\mathbf{CP}^1$s that intersect transversely at one point	$\begin{pmatrix} 0 & -1 \\ 1 & -1 \end{pmatrix}$	4
I_0^* $(\widetilde{D}_4)$	diagram with central vertex 2 and four vertices 1	$\begin{pmatrix} -1 & 0 \\ 0 & -1 \end{pmatrix}$	6
I_b^* $(\widetilde{D}_{4+b})$	diagram with $b+1$ central vertices labeled 2 and four end vertices labeled 1	$\begin{pmatrix} -1 & 0 \\ -b & -1 \end{pmatrix}$	$b+6$
II^* $(\widetilde{E}_8)$	diagram 1–2–3–4–5–6–4–2 with a 3 branching from the 6	$\begin{pmatrix} 0 & 1 \\ -1 & 1 \end{pmatrix}$	10
III^* $(\widetilde{E}_7)$	diagram 1–2–3–4–3–2–1 with a 2 branching from the central 4	$\begin{pmatrix} 0 & 1 \\ -1 & 0 \end{pmatrix}$	9
IV^* $(\widetilde{E}_6)$	diagram 1–2–3–2–1 with a branch 2–1 from the central 3	$\begin{pmatrix} -1 & 1 \\ -1 & 0 \end{pmatrix}$	8

Each vertex corresponds to $\mathbf{CP}^1$, its suffix means the divisor's multiplicity, and two divisors intersect each other transversely in one point if and only if the corresponding vertices are connected by an edge. Here we note that the monodromy representations in Table 1 are different from those in [K1, BPV], but this is due to a different choice of the bases and the directions of the monodromies we used. We also note that $_mI_b$ ($m \geq 2$) are the only multiple fibers. The fiber of type $_1I_b$ is denoted by I_b.

DEFINITION (1-6). The fiber of type $_mI_0$ is called a *multiple torus of multiplicity* m.

Sometimes framed links are useful for representing the regular neighborhoods of these singular fibers. Some of them are written in Figure 1 (see p. 100). (The other cases are also easy to write down. See [HKK].) A framed link represents a handle decomposition of a 4-manifold such that the link component with suffix n corresponds to a 2-handle with framing n, and a dotted component corresponds to a dual 1-handle obtained from D^4 by removing a tubular neighborhood of a 2-disk in D^4 bounded by that component (see [Ki, Man1] for the details). Now we will describe the structure of the fibration $\pi : X \to B$ in detail. To this end, let F_i ($i = 1, \ldots, s$) be the multiple tori for π (in the sense of Definition (1-6)) and F'_j ($j = 1, \ldots, t$) the other singular fibers in X. Put $p_i = \pi(F_i)$, $p'_j = \pi(F'_j)$. Take another point p_0 in B_0 for convenience and put $F_0 = \pi^{-1}(p_0)$. Furthermore, let Δ_i and Δ'_j be the small disk neighborhoods of p_i and p'_j respectively. Put

$$B_0 = \mathrm{Cl}(B - \bigcup_{i=0}^{s} \Delta_i - \bigcup_{j=1}^{t} \Delta'_j) \qquad \text{(Cl means the closure)}$$

and take the loops $\overline{\alpha}_i$, $\overline{\beta}_i$ ($i = 1, \ldots, g$) which represent the standard generators of $H_1(B, \mathbf{Z})$ and $\overline{q}_i = \partial \Delta_i$, $\overline{q}'_j = \partial \Delta'_j$ connected to $*$ by the paths as in Figure 2 (see p. 100). Then the description (1)-(4) of X above is replaced by the more detailed information that follows.

(i′) $\pi : \pi^{-1}(B_0) \to B_0$. This is a T^2-bundle over B_0 with cross section $\widetilde{B}_0$ (such a cross section exists even if $s = t = 0$ since the disk Δ_0 is removed). If the monodromy representations of $\overline{\alpha}_i$, $\overline{\beta}_i$, $\overline{q}_i$, $\overline{q}'_j$ with respect to the basis (ℓ, h) of $\pi^{-1}(*)$ are given by A_i, B_i, P_i, $P'_j \in \mathrm{SL}(2, \mathbf{Z})$ respectively, then

(∗) $\qquad P_i = I, \quad \prod_{i=1}^{g}[A_i, B_i]\prod_{j=1}^{t} P'_j = I \quad$ (I is the identity matrix),

(∗∗) $\qquad P'_i = Q_i P''_i Q_i^{-1}, \quad Q_i \in \mathrm{SL}(2, \mathbf{Z})$

where P''_i is the monodromy representation of F'_i in Table 1 and we define $[A_i, B_i] = A_i B_i A_i^{-1} B_i^{-1}$. Moreover, putting

(∗∗∗) $\qquad \begin{aligned} q_i &= \widetilde{B}_0 \cap \pi^{-1}(\overline{q}_i), & q'_j &= \widetilde{B}_0 \cap \pi^{-1}(\overline{q}'_j), \\ \alpha_i &= \widetilde{B}_0 \cap \pi^{-1}(\overline{\alpha}_i), & \beta_i &= \widetilde{B}_0 \cap \pi^{-1}(\overline{\beta}_i), \end{aligned}$

we have the following relations in the fundamental group of $\pi^{-1}(B_0)$:

$N(I_1)$ O $N(II)$ O

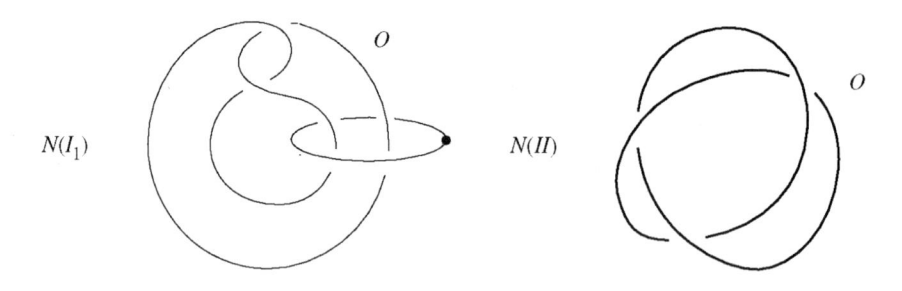

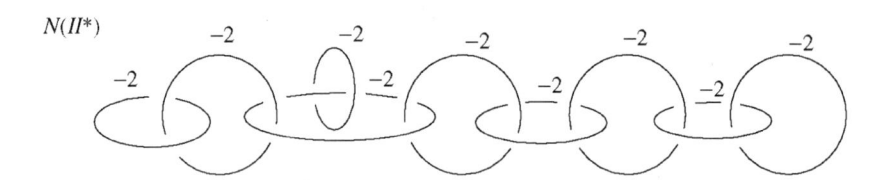

$N(II^*)$

FIGURE 1

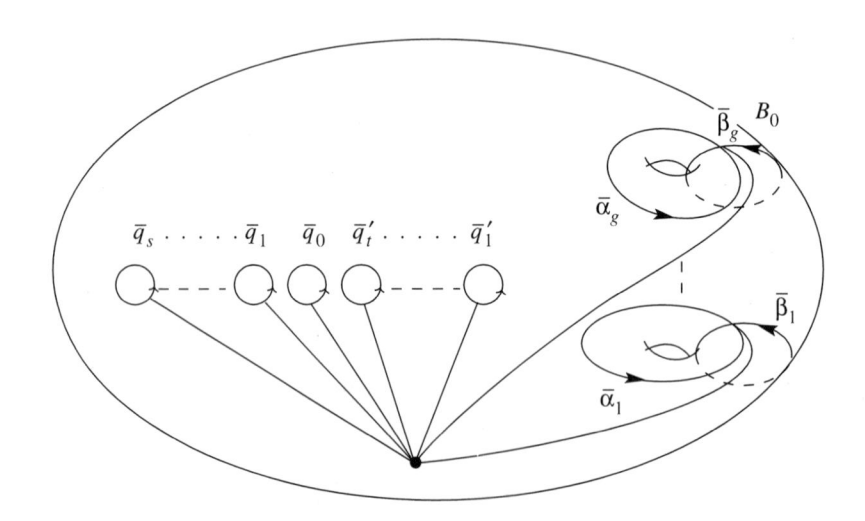

FIGURE 2

$$(\alpha_i \ell \alpha_i^{-1}, \alpha_i h \alpha_i^{-1}) = (\ell, h) A_i, \qquad (\beta_i \ell \beta_i^{-1}, \beta_i h \beta_i^{-1}) = (\ell, h) B_i,$$
$$(q_i \ell q_i^{-1}, q_i h q_i^{-1}) = (\ell, h), \qquad (q_j' \ell q_j'^{-1}, q_j' h q_j'^{-1}) = (\ell, h) P_j'.$$

Here, since ℓ and h are commutative in the fundamental group, $\ell^a h^b$ is written as $a\ell + bh$ and on the right-hand sides above the matrices act naturally from the right.

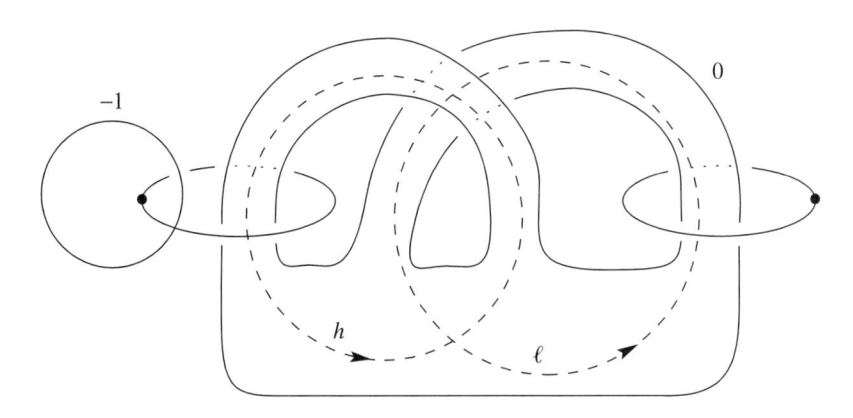

FIGURE 3

REMARK. By an appropriate change of $\widetilde{B}_0$ we can replace a given q_i or q'_j $(i, j \geq 1)$ by another loop representing $q_i \ell^a h^b$ or $q'_j \ell^{a'} h^{b'}$ $(a, b, a', b' \in \mathbf{Z})$ in the fundamental group. By the influence on $\pi^{-1}(\overline{q}_0)$ of this process q_0 is replaced by another one.

(ii') How to glue the singular fiber $\pi : N(F'_j) \to \Delta'_j$. The attaching map $\varphi_j : \partial N(F'_j) \to \pi^{-1}(\overline{q}'_j) \subset \pi^{-1}(B_0)$ induces the transformation corresponding to Q_j in (i'), (**) on the fiber. Furthermore, if F'_j is not a multiple fiber, choose $\widetilde{B}_0$ so that $\widetilde{B}_0$ and the lift $\widetilde{\Delta}'_j$ of $\pi : N(F'_j) \to \Delta'_j$ are patched together (see the remark above).

As is seen in §4 the fibers of type I_1 and $_m I_0$ are most important for determining the diffeomorphism type of X. So we will explain how these fibers (and those of type $_m I_b$) are attached. See (4–7) in §4 for the other fibers.

(iii') How to attach I_1. The neighborhood $N(I_1)$ of the fiber of type I_1 is diffeomorphic to a sum of $B^2 \times T^2$ and a 2-handle h_2. Here $* \times T^2 \subset \partial B^2 \times T^2$ corresponds to the general fiber in $N(I_1)$, and h_2 is attached along a simple closed curve $C = r\ell + sh$ $(r, s \in \mathbf{Z}, \gcd(r, s) = 1)$ on $* \times T^2$ (we fix the base (ℓ, h)) with framing -1. The diffeomorphism type of $N(I_1)$ does not depend on the choice of C, but its monodromy representation depends on C. For example, if we take $C = h$, then the monodromy representation of I_1 with respect to (ℓ, h) is $\begin{pmatrix} 1 & 0 \\ 1 & 1 \end{pmatrix}$ and $N(I_1)$ is represented by the following framed link (it is not difficult to transform this to the framed link in Figure 1).

In Figure 3 a disk which is a core of the 2-handle with framing -1 is attached to "a vanishing cycle" and passes through the singularity of I_1. Furthermore, for the given monodromy representation of I_1 (which is necessarily the conjugate of $\begin{pmatrix} 1 & 0 \\ 1 & 1 \end{pmatrix}$) the way of attaching $N(I_1)$ is determined as follows:

(RI) h_2 is attached along $r\ell + sh$ $\iff$ the monodromy representation of I_1 with respect to (ℓ, h) is given by

$$P \begin{pmatrix} 1 & 0 \\ 1 & 1 \end{pmatrix} P^{-1} = \begin{pmatrix} rs + 1 & -r^2 \\ s^2 & -rs + 1 \end{pmatrix}, \quad P = \begin{pmatrix} p & r \\ q & s \end{pmatrix} \in \mathrm{SL}(2, \mathbf{Z})$$

(iv') How to attach multiple tori. For any fiber $F = {}_m I_0$ take a basis (ℓ, h) of the general fiber of $\pi : \partial N(F) \to \partial \Delta$ so that it coincides with that of $\pi^{-1}(*)$ in (i'),

and take a lift q of $\partial \Delta$ as in (i'), (∗∗∗). On the other hand, since $N(F) \cong B^2 \times T^2$, we can take
$$Q = \partial B^2 \times * \times *, \quad L = * \times S^1 \times *, \quad H = * \times * \times S^1.$$

Then (L, H, Q) and (ℓ, h, q) are the bases of $H_1(\partial N(F), \mathbf{Z})$, and we can assume that the attaching map $\varphi : \partial N(F) \to \pi^{-1}(\partial \Delta) \subset \pi^{-1}(B_0)$ is linear with respect to these bases. So φ is represented by the matrix defined by

(RII) $\qquad (L, H, Q) = (\ell, h, q)A, \quad A = \begin{pmatrix} * & * & a \\ * & * & b \\ * & * & m \end{pmatrix} \in \mathrm{SL}(3, \mathbf{Z}).$

In fact, we only need (m, a, b) to determine how to attach $N(F)$ (the difference of the other entries is absorbed by some self-diffeomorphisms of $B^2 \times T^2$).

DEFINITION (1-7). A multiple torus attached according to (RII) is said to be of type (m, a, b) with respect to (q, ℓ, h). This definition is valid also in the case $m = 1$.

If $m \geq 2$, the fibration $\pi : N(F) \to \Delta$ is given by the following model. Define a $\mathbf{Z}_m$-action on $B^2 \times T^2 = \{(z, \theta, \varphi) \mid z \in \mathbf{C}, |z| \leq 1, \theta, \varphi \in \mathbf{R} \bmod 1\}$ by
$$\rho(z, \theta, \varphi) = (\exp(2\pi i/m)z, \theta - a/m, \varphi - b/m)$$
for a generator ρ of $\mathbf{Z}_m$. Then the projection from $B^2 \times T^2$ to the first factor induces the map $\pi' : (B^2 \times T^2)/\mathbf{Z}_m \to B^2/\mathbf{Z}_m$. This is naturally identified with $\pi : N(F) \to \Delta$. Under this identification q corresponds to the image into $(B^2 \times T^2)/\mathbf{Z}_m$ of $\{\exp(2\pi i t/m)z, \theta - ta/m, \varphi - tb/m) \mid 0 \leq t \leq 1\}$.

(v') How to attach $_mI_b$ ($m \geq 1, b \geq 1$). To describe $\pi : N(F) \to \Delta$ for $F = {}_mI_b$ we start with the projection $\tilde{\pi} : N(\tilde{F}) \to \tilde{\Delta}$ for $\tilde{F} = I_{mb}$. $N(\tilde{F})$ is diffeomorphic to a union of $B^3 \times S^1$ and mb 2-handles attached along its boundary (with framing -2). The union of the part with oblique lines in Figure 4 and the disks parallel to the cores of 2-handles corresponds to the fiber in $\partial N(\tilde{F})$.

On the other hand, the positions of mb 2-handles can be adjusted so that we can define the orientation-preserving $\mathbf{Z}_m$-action on $N(I_{mb})$ whose generator induces the rotation through angle $-2\pi a/m$ on the S^1-factor ($\gcd(m, a) = 1$) and induces the rotation through angle $2\pi/m$ on $\tilde{\Delta}$. Hence we get the projection $\pi : N(I_{mb})/\mathbf{Z}_m \to \tilde{\Delta}/\mathbf{Z}_m$ induced by $\tilde{\pi}$, which is the desired model of $\pi : N(F) \to \Delta$.

DEFINITION (1-8). If a regular neighborhood $N(F_0)$ of the general fiber F_0 for the singular torus fibration $\pi : X \to B$ (including the case of elliptic surfaces) is removed from X and the neighborhood of some multiple torus $_mI_0$ is attached according to (RII) in (iv'), then we get a new fibration $\pi' : X' \to B$. This process is called a *logarithmic transformation of multiplicity m*. Similarly, the process of replacing the neighborhood $N(\tilde{F})$ of the fiber $\tilde{F}$ of type I_b by the neighborhood $N(F)$ of the fiber F of type $_mI_b$ according to the model in (v') is also called a logarithmic transformation of multiplicity m.

If X is an elliptic surface, the above logarithmic transformation is performed so that the new manifold X' is also an elliptic surface (that is, X' also admits a complex structure) for any (m, a, b) ($\gcd(m, a, b) = 1$) or (m, a) ($\gcd(m, a) = 1$) [**K1, BPV**]. In fact, any elliptic surface is obtained from one without multiple fibers by such transformations. We will see in §4 that only m in the data (m, a) for the attaching of $_mI_b$ ($b \geq 1$) in (v') is essential (that is, if we use another a for the regluing process we get the same manifold up to diffeomorphism). (The situation is different for $_mI_0$.)

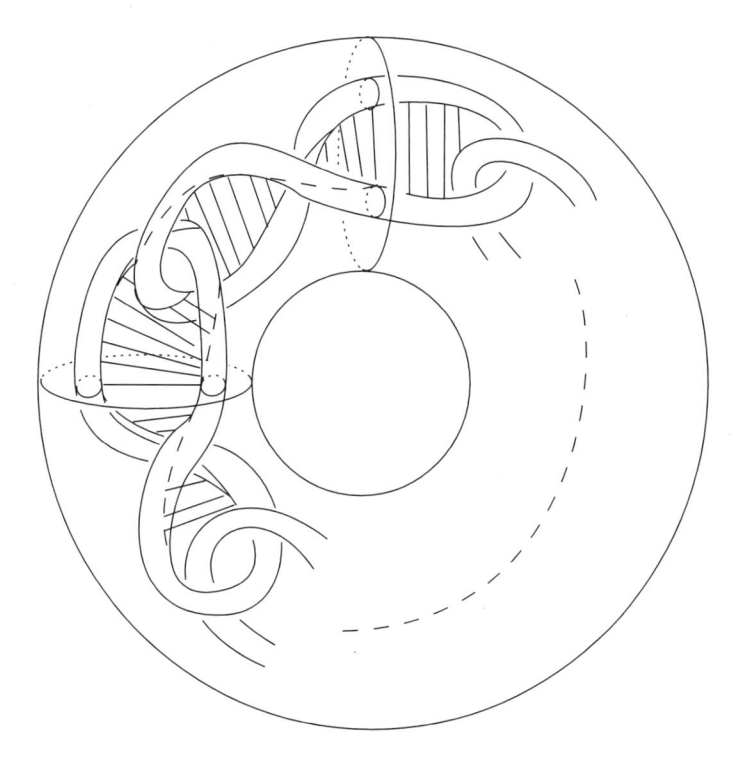

FIGURE 4. $B^2 \times S^1 \subset \partial B^3 \times S^1$

§2. Basic properties of elliptic surfaces

In this section, we will discuss the basic properties of elliptic surfaces $\pi: X \to B$ (including the ith Chern class $c_i(X)$ and the second Stiefel-Whitney class $w_2(X)$) and their positions in the class of the complex surfaces (see [**K1, K2, BPV**] for the details). Let K_X be the canonical divisor of X, F_i ($i = 1, \ldots, s$) the multiple fibers contained in X with multiplicity m_i, and g the genus of B.

(2-1). (i) $c_2(X) = e(X) = \sum e(\text{the singular fiber of } X) = 12(1 - q + p_g)$ where $q = \dim H^1(X, \mathcal{O}_X)$, $p_g = \dim H^2(X, \mathcal{O}_X)$, and $\mathcal{O}_X$ is the structure sheaf of X.
(ii) $K_X = \pi^*\mathscr{L} + \sum_{i=1}^{s}(m_i - 1)F_i$ where $\mathscr{L}$ is the divisor on B with $\deg \mathscr{L} = (1 - q + p_g) - 2(1 - g)$.
(iii) $c_1(X) = -c_1(K_X)$, $w_2(X) \equiv c_1(X) \pmod{2}$, $c_1^2(X) = 0$.
(iv) $\sigma(X) = (c_1^2(X) - 2c_2(X))/3 = -2e(X)/3$.
(v) $b_2^+(X) = 1 + 2p_g$.

From these we deduce

(2-2) $\qquad e(X) = \begin{cases} 0 & \text{if } X \text{ has no singular fibers} \\ & \text{other than multiple tori} \\ \text{positive} & \text{otherwise.} \end{cases}$

(This is also valid for any singular torus fibration.)

(2-3). If $e(X) > 0$, then $b_1(X) = 2q = 2g$ and $\deg \mathscr{L} = p_g + q - 1$.

In general, for any complex surface X a number called the Kodaira dimension $\text{kod} X$ (with value $-\infty, 0, 1,$ or 2) is defined (for the definition see [BPV] for example). Now we will give the list of all complex surfaces that can be (minimal) elliptic surfaces together with their properties according to [BPV] using the Kodaira dimensions. (More explicit descriptions will be given in §§4,5.)

The cases with $\text{kod} X = -\infty$.

(a) Rational surfaces: Among these nine points the blow-up of $\mathbf{CP}^2$, i.e., $\mathbf{CP}^2 \sharp 9\overline{\mathbf{CP}}^2$ ($e = 12$, $b_1 = 0$) is the only one that can be an elliptic surface.

(b) Ruled surfaces of genus 1: These are $\mathbf{CP}^1$-bundles over the Riemann surfaces of genus 1 and there are just two diffeomorphism types among them according as $w_2 = 0$ or $w_2 \neq 0$.

(c) Hopf surfaces: These are the surfaces whose universal covers are $\mathbf{C}^2 - \{0\}$. In fact, it is sufficient for the surfaces to be Hopf surfaces that their universal covers are homeomorphic to $S^3 \times \mathbf{R}$ [K1, K2]. In [BPV] the surface X with $\text{kod} X = -\infty$ and $b_1(X) = 1$ is said to be of class VII. The surface of class VII is homeomorphic to an elliptic surface only if it is a Hopf surface. Notice the following theorem concerning this claim.

(2-4), [B, LYZ]. *Any surface of type* VII *with* $b_2 = 0$ *is either a Hopf surface or an Inoue surface.*

See [In, N] for the Inoue surfaces. None of them is homeomorphic to an elliptic surface.

The cases when $\text{kod} X = 0$. The cases (d)–(i) below cover this class.

(d) $K3$ surfaces: These are surfaces such that $b_1(X) = 0$ and K_X is trivial ($e(X) = 24$). They are all diffeomorphic [K1] and so we will denote them simply by $K3$ hereafter. $K3$ is simply connected and its intersection form is $2E_8 \oplus 3U$.

(e) Enriques surfaces: These are surfaces such that K_X is nontrivial but $K_X^{\otimes 2}$ is trivial. The diffeomorphism type of these is unique, their fundamental groups are $\mathbf{Z}_2$, and their intersection forms on $H_2/\text{Torsion}$ are $E_8 \oplus U$ (but $w_2 \neq 0$). Their universal covers are $K3$ [BPV].

(f) Hyperelliptic surfaces: These are elliptic curve bundles over elliptic curves with $b_1 = 2$ ($e = 0$). There are just seven such surfaces.

(g) Primary Kodaira surfaces: These are elliptic curve bundles over elliptic curves with $b_1 = 3$ ($e = 0$).

(h) Secondary Kodaira surfaces: These are elliptic surfaces over $\mathbf{CP}^1$ with $b_1 = 1$ whose universal coverings are in (g) ($e = 0$).

(i) Complex tori of dimension two: ($e = 0, b_1 = 4$).

The cases when $\text{kod} X = 1$. (All in this class are elliptic surfaces.)

(j) Properly elliptic surfaces: These are elliptic surfaces other than those appearing in (a)–(i). (If $e > 0$, then $b_1 = 2g$, and in the case where $e = 0$ we have either $b_1 = 2g + 2$ or $2g + 1$ where g is the genus of the base B.)

These examples are elliptic surfaces only when they are equipped with specific complex structures in general. Furthermore, every example with $b_1 \equiv 0$ (mod 2) is diffeomorphic to an algebraic surface.

According to the results that will be discussed in §4, the diffeomorphism type of any elliptic surface $\pi : X \to B$ with $e(X) > 0$ and $B = \mathbf{CP}^1$ is determined by $e(X) = 12(1 + p_g)$ (hereafter we will write $k = 1 + p_g$) and the set of multiplicities of the multiple fibers $(m_1, \ldots, m_s)$. Hence we denote such X by $S_k(m_1, \ldots, m_s)$. Considering $S_k(1, m_2, \ldots, m_s)$ as $S_k(m_2, \ldots, m_s)$, for example, we can see that such representations are also valid if $m_i = 1$. Some of them have special names as follows.

$S_1 \cong S_1(m)$. A rational elliptic surface $\cong \mathbf{CP}^2 \sharp 9\overline{\mathbf{CP}}^2$ (not depending on m).

$S_1(m_1, m_2)$ $(m_1, m_2 \geq 2, \gcd(m_1, m_2) = 1)$. Dolgachev surfaces (sometimes the surfaces with $\gcd(m_1, m_2) \geq 2$ are also called by the same names). Any of them is homeomorphic, but not diffeomorphic, to S_1.

S_2. The $K3$ surface.

$S_2(m_1, m_2)$ $(m_1 \equiv m_2 \equiv 1 \pmod{2}, \gcd(m_1, m_2) = 1$, either $m_1 > 1$ or $m_2 > 1)$. Such a surface is homotopy equivalent to $K3$ (actually homeomorphic to $K3$) and so is called a homotopy $K3$ surface [K3]. It is proved that none of them is diffeomorphic to $K3$ [FM2, Mon].

$S_1(2, 2)$. The Enriques surface.

§3. Remarks on Seifert fibered 3-manifolds

In this section we will summarize the results about 2-orbifolds and Seifert fibered 3-manifolds (here we only consider the closed oriented ones) which are needed later. First, a singular 2-manifold whose singularities have neighborhoods of the form $B^2/\mathbf{Z}_m$ (where $\mathbf{Z}_m$ acts as rotations) is called a *2-orbifold* (V-manifold of dimension two) [T, Sc]. The 2-orbifold B is determined by its genus g and the set of multiplicities $(m_1, \ldots, m_k)$ of the singular points on B (which correspond to the above m). We denote B by $\Sigma_g(m_1, \ldots, m_k)$ (or $S^2(m_1, \ldots, m_k)$ if $g = 0$). They are divided into the following four classes:

(1) B "bad". $S^2(m), S^2(m, n)$, for $m \neq n, m, n \geq 2$;

(2) B spherical. S^2, $S^2(n, n)$, $S^2(2, 2, n)$ for $n \geq 2$, $S^2(2, 3, 3)$, $S^2(2, 3, 4)$, $S^2(2, 3, 5)$;

(3) B euclidean. $S^2(2, 2, 2, 2)$, $S^2(2, 3, 6)$, $S^2(2, 4, 4)$, $S^2(3, 3, 3)$, $T^2 \cong \Sigma_1$;

(4) B hyperbolic. Those except for (1)–(3).

If B belongs to (2), (3), or (4), then we can write $B = X/\Gamma$ for $X = S^2$, E^2 (the euclidean plane), or H^2 (the upper half-plane) respectively. Here Γ is a discrete subgroup of orientation-preserving isometries on X. The group Γ is called an *orbifold fundamental group of B* and is denoted by $\pi_1^{\text{orb}} B$ [Sc]. If B is bad, B cannot be written in the form X/Γ, but we can put $\pi_1^{\text{orb}} S^2(m, n) \cong \mathbf{Z}/\gcd(m, n)\mathbf{Z}$ (where $S^2(n, 1) = S^2(n)$).

(3-1). *If $B \neq S^2(m, n)$, then $\pi_1^{\text{orb}} B$ is not cyclic. Furthermore, in this case the isomorphism type of B is determined by $\pi_1^{\text{orb}} B$.*

This fact can be proved combinatorially [ZVC] and also geometrically [N, Wo]. In the case where B is spherical, we can prove this by checking $\pi_1^{\text{orb}} B$ directly.

Secondly, a closed oriented 3-manifold M is called a *Seifert fibered 3-manifold* if it admits a projection $\pi : M \to B$ to an oriented closed surface B such that its general fiber is S^1 [Or]. In this case any singular fiber is a multiple fiber on which a general fiber covers with a certain multiplicity (say m) and the way of attaching its neighborhood $\cong S^1 \times B^2$ is given by the type (m, a) $(\gcd(m, a) = 1)$. The definition of the type is parallel to that for the multiple torus in §1 and is also valid even if $m = 1$ as in §1. Then M is determined by the genus g of B and the types of the multiple fibers as follows. These are called the *Seifert invariants of M* (the representation here is according to [NR]):

$$\{g,(1,b),(m_1,a_1),\ldots,(m_k,a_k)\}.$$

Here it is natural to consider B as the orbifold $\Sigma_g(m_1,\ldots,m_k)$. Now we will describe examples of Seifert fibered 3-manifolds appearing in the next section and fix their notation (there are some overlaps between (1) and (2) below).

(1) Brieskorn homology 3-spheres.
We call $\Sigma(p,q,r) = \{(z_1,z_2,z_3) \in \mathbf{C}^3 \mid \sum |z_i|^2 = \varepsilon > 0, z_1^p + z_2^q + z_3^r = 0\}$ a *Brieskorn manifold of type p,q,r*. This is a Seifert fibered 3-manifold and, in particular, if p, q, r are pairwise coprime, then it is a homology 3-sphere ($H_*(\Sigma(p,q,r),\mathbf{Z}) = H_*(S^3,\mathbf{Z})$). This is given the orientation induced from that of $\mathbf{C}^3$.

(2) Spherical 3-manifolds S^3/Γ (where Γ is a finite subgroup of SO(4) acting freely on S^3). Every such manifold is contained in the following list.
   (a) Lens spaces $L(p,q)$ (including $S^3 = L(1,0)$). Each of them has infinitely many fibrations over the base of the form $S^2(m,n)$. See [**Or**] for the detailed representations. Here we only remark that if $q^2 \equiv 1 \pmod{p}$, then it corresponds to the Seifert fibrations of the following forms (hereafter the genus of the base is omitted in the representations since it is always 0):

(L) $$L(p,q) \cong \{(1,b),(\alpha,\beta),(\alpha,\beta)\}$$

where $p = \alpha(2\beta + \alpha b)$, $q = m\alpha + n\beta$ for the intergers m, n satisfying $m\alpha - n(\beta + \alpha b) = 1$.

   (b) $F_1 = \{(1,b),(2,1),(2,1),(n,k)\}$, $b \in \mathbf{Z}$, $n \geq 2$.
   In particular, we denote the case with $(n,k) = (2,1)$ by F_1'.
   (c) $F_2 = \{(1,b),(2,1),(3,1),(3,k)\}$, $k = \pm 1$, $b \in \mathbf{Z}$.
   (d) $F_3 = \{(1,b),(2,1),(3,1),(4,k)\}$, $k = \pm 1$, $b \in \mathbf{Z}$.
   (e) $F_4 = \{(1,b),(2,1),(3,1),(5,k)\}$, $k = \pm 1, \pm 2$, $b \in \mathbf{Z}$.
   (Each of these classes contains infinitely many examples according to the change of $b \in \mathbf{Z}$ and k but for simplicity we denote them by the same symbol here.)

Next we will define (orientation-preserving) self-diffeomorphisms σ, ρ on some spherical 3-manifolds that are needed in §5. Each of them is well-determined up to isotopy by the following definitions.
   (1) σ. This is a fiber-preserving self-diffeomorphism of order two on either $L(p,q) = \{(1,b),(\alpha,\beta),(\alpha,\beta)\}$ (which satisfies (L)), F_1, or F_2. In each case, σ induces an automorphism on the base orbifold which is a rotation interchanging the singular points of the same multiplicities and fixing the remaining singular point (but σ on F_2 is defined only when $k = 1$).
   (2) ρ. This is a fiber-preserving self-diffeomorphism of order three on F_1'. The automorphism on $S^2(2,2,2)$ induced by ρ is a rotation that permutes the three singular points cyclically.

§4. Decompositions of the singular fibers and the topology of elliptic surfaces with positive Euler numbers (I)

If we are concerned with the diffeomorphism type of the singular torus fibration $\pi : X \to B$ (including an elliptic surface), the following theorem of Moishezon is quite important. It asserts that every singular fiber is decomposed as a sum of fibers of type $I_1 = {}_1I_1$ and ${}_mI_0$ ($m \geq 2$). Here we give it in a simplified form from the topological point of view.

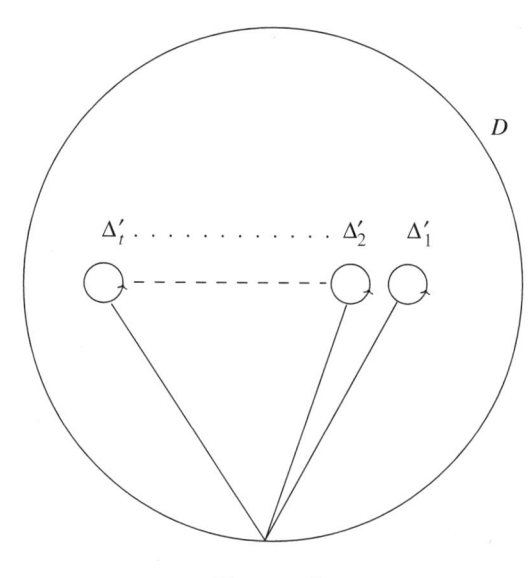

FIGURE 5

(4-1), [**M1**]. (1). *For a fiber F of type $_mI_b$ ($m \geq 2$, $b > 0$) there is a singular torus fibration $\pi' : N' \to B^2$ over B^2 that contains just one fiber of type I_b and just one fiber of type $_mI_0$ as its singular fibers such that N' is diffeomorphic to the neighborhood $N(F)$ of F (by a diffeomorphism inducing the identity on the boundaries).*

(2). *For every nonmultiple singular fiber F there exists a singular torus fibration $\pi'' : N'' \to B^2$ over B^2 that contains only fibers of type I_1 as its singular fibers such that $N(F)$ and N'' are diffeomorphic by a diffeomorphism inducing the identity on the boundaries.*

DEFINITION (4-2). We call a singular torus fibration a *Lefschetz fibration* if it contains only the fibers of type I_1 and $_mI_0$ as its singular fibers.

(4-1) shows that any singular torus fibration is diffeomorphic to a Lefschetz fibration.

DEFINITION (4-3). Consider the Lefschetz fibration X over the disk D without multiple fibers such that the neighborhoods Δ'_i of the images of the fibers F'_i of type I_1 ($i = 1, \ldots, t$) are located as in Figure 5. If the monodromy representation of F_i with respect to the basis (ℓ, h) of the fiber over the base point $*$ on ∂D is $P_i \in \mathrm{SL}(2, \mathbf{Z})$, then we write $X = X(P_1, \ldots, P_t)$. This space admits a cross section $\widetilde{D}$ for the base space.

Since P_i is conjugate to $\begin{pmatrix} 1 & 0 \\ 1 & 1 \end{pmatrix}$, the fibration $\pi : X(P_1, \ldots, P_t) \to D$ is uniquely determined according to the rule §1 (RI) once (ℓ, h) and P_i are fixed. For a Lefschetz fibration $\pi : X \to B$ with $e(X) > 0$ take a disk D in B so that it contains all Δ'_i ($i = 1, \ldots, t$) and does not contain any Δ_j ($j = 0, \ldots, s$) (the base point is taken on ∂D). Put $B'_0 = B - \bigcup \Delta_j - D$; then the fibration of X is described as the following information in place of (i′)–(v′) in §1.

(0) $X \cong \pi^{-1}(B_0') \cup_{\varphi'} X(P_1, \ldots, P_t) \cup (\bigcup_{\varphi_i} N(F_i))$

(1) $\pi : \pi^{-1}(B_0') \to B_0'$ is a T^2-bundle over B_0' admitting the cross section $\widetilde{B}_0'$ whose monodromy representations of $\overline{\alpha}_i$, $\overline{\beta}_i$, $\overline{q}_i$, ∂D with respect to the basis (ℓ, h) of the fiber over the base point are A_i, B_i, I, $\prod_{j=1}^{t} P_j$. Here we have $\prod [A_i, B_i] \prod P_j = I$.

(2) The map $\varphi' : \partial X(P_1, \ldots, P_t) \to \pi^{-1}(\partial D) \subset \pi^{-1}(B_0')$ is a fiber-preserving diffeomorphism that maps $\partial \widetilde{D}$ to $\partial \widetilde{B}_0' \cap \pi^{-1}(\partial D)$ where $\widetilde{D}$ is a cross section of $\pi : X(P_1, \ldots, P_t) \to D$. (Accordingly there exists a cross section of $B - \bigcup \Delta_j$ with respect to π.)

(3) $N(F_j)$ is a neighborhood of a multiple torus of type (m_j, a_j, b_j) with respect to (q_j, ℓ, h), where $q_j = \widetilde{B}_0' \cap \pi^{-1}(\partial \Delta_j)$ (we assume $m_0 = 1$).

On the other hand, it is known that the monodromy representations for $X(P_1, \ldots, P_t)$ are transformed by pull-backs by some automorphisms of the base space as follows. (They are called elementary transformations.)

(4-4), [M1].

$$X(P_1, \ldots, P_i, P_{i+1}, \ldots, P_t) \cong X(P_1, \ldots, P_{i+1}, P_{i+1}^{-1} P_i P_{i+1}, \ldots, P_t),$$
$$X(P_1, \ldots, P_i, P_{i+1}, \ldots, P_t) \cong X(P_1, \ldots, P_i P_{i+1} P_i^{-1}, P_i, \ldots, P_t).$$

Both of these relations are induced by diffeomorphisms inducing the identity on the boundaries.

Using this, Moishezon proved the following.

(4-5), [M1]. *For any Lefschetz fibration $X(P_1, \ldots, P_t)$ over B^2 with $\prod_{i=1}^{t} P_i = I$ (that is, $\partial X(P_1, \ldots, P_t) \cong T^3$), we have $X(P_1, \ldots, P_t) \cong X(P_1', \ldots, P_t')$ by a diffeomorphism inducing the identity on the boundaries. Here $t = 12k$ for some k and*

$$P_i' = \begin{cases} s_1 & \text{if } i \text{ is odd,} \\ s_2 & \text{if } i \text{ is even.} \end{cases}$$

Here $s_1 = \begin{pmatrix} 1 & 0 \\ 1 & 1 \end{pmatrix}$, $s_2 = \begin{pmatrix} 1 & -1 \\ 0 & 1 \end{pmatrix}$. *In particular, $e(X(P_1, \ldots, P_t)) = 12k$.*

Similar claims hold also for the singular fibers as follows. In either case below, the diffeomorphism is realized by that inducing the identity on the boundaries (whose representation is based on (ℓ, h) giving the monodromy representations in §1, Table 1).

(4-6)
$$N(II) \cong X(s_1, s_2), \quad N(III) \cong X(s_1, s_2, s_1) \cong X(s_2, s_1, s_2),$$
$$N(IV) \cong X(s_1, s_2, s_1, s_2),$$
$$N(II^*) \cong X(s_1, s_2, s_1, s_2, s_1, s_2, s_1, s_2, s_1, s_2),$$
$$N(III^*) \cong X(s_1, s_2, s_1, s_2, s_1, s_2, s_1, s_2, s_1),$$
$$N(IV^*) \cong X(s_1, s_2, s_1, s_2, s_1, s_2, s_1, s_2),$$
$$N(I_b^*) \cong X(s_1, s_2, s_1, s_2, s_1, s_2, \overbrace{s_1, \ldots, s_1}^{b}) \quad (b \geq 0),$$
$$N(I_b) \cong X(\overbrace{s_1, \ldots, s_1}^{b}) \quad (b \geq 1).$$

These claims are also proved by the transformations of the framed links if the numbers of the divisors are small [HKK, Mt1, U1]. Furthermore, from (4-6), the framed link pictures of $N(F)$, and the observations about the self-diffeomorphisms of $\partial N(F)$ we deduce the following.

(4-7). *For any singular fiber F other than ${}_mI_b$ ($m \geq 1$, $b \geq 0$), any orientation-preserving self-diffeomorphism of $\partial N(F)$ extends to that of $N(F)$. Hence if we reglue such an $N(F)$ in any way the diffeomorphism type of the resulting manifold is the same as the old one.*

For a Lefschetz fibration X (with $e(X) > 0$) over a closed surface B, Matsumoto [**Mt2**] shows the following.

(4-8), [**Mt2**]. *If $e(X) > 0$, then there exists a Lefschetz fibration $\pi' : X' \to B$ whose monodromies of $\bar{\alpha}_i$, $\bar{\beta}_i$ ($i = 1, \ldots, g$) in Figure 2 are all trivial ($A_i = B_i = I$) so that $X \cong X'$ by a diffeomorphism that is the identity near the multiple tori.*

By (4-5) and (4-8) we can assume in the previous construction (0)–(3) of X that
$$X(P_1, \ldots, P_t) = X(\overbrace{s_1, s_2, \ldots, s_1, s_2}^{12k}), \quad \pi^{-1}(B'_0) \cong B'_0 \times T^2$$
(in this case $e(X) = 12k$). Concerning the attaching of the multiple tori we have

(4-9). *Put $B_\natural = B - \bigcup(\text{the image of the singular fiber})$. Then the monodromy representation $\pi_1 B_\natural \to \mathrm{SL}(2, \mathbf{Z})$ along the loops on $B_\natural$ is surjective.*

This comes from the fact that s_1 and s_2 generate $\mathrm{SL}(2, \mathbf{Z})$. Using (4-9) we can see the following [**M1, U2, G1**].

(4-10). (1) *For any Lefschetz fibration $\pi : X \to B$ with $e(X) > 0$, if we remove a neighborhood $N(F_0)$ of a torus of type $(1, a_0, b_0)$ from X and reglue it as a torus of type $(1, 0, 0)$, then the resulting manifold is diffeomorphic to X.*

(2) *If a neighborhood $N(F_i)$ of a multiple torus F_i of type (m_i, a_i, b_i) for any $i \geq 1$ is removed from the same X and is replaced by a neighborhood of a new multiple torus of type $(m_i, 1, 0)$ then the resulting manifold is diffeomorphic to X.*

In particular, the matrices in §1, (RII) which determine how to attach $N(F_i)$s can be taken so that they are of the following form simultaneously:
$$I \quad (\text{for } m_0 = 1), \quad \begin{pmatrix} 1 & 0 & 0 \\ 0 & 0 & 1 \\ 0 & -1 & m_i \end{pmatrix} \quad (\text{for } m_i \geq 2).$$

These results lead us to the following conclusions.

(4-11). *The diffeomorphism type of the singular torus fibration X over a closed surface B with $e(X) > 0$ is uniquely determined by $e(X)$ (which is necessarily a multiple of 12), the genus g of B, and the set of multiplicities $(m_1, \ldots, m_s)$ of the multiple tori.*

(4-12), [**Mt2**]. *The diffeomorphism type of any elliptic surface X without multiple fibers and with $e(X) > 0$ is determined by $e(X)$ and the genus g of the base space B.*

(4-13). *Every elliptic surface X without multiple fibers and with $e(X) > 0$ has a smooth cross section $\widetilde{B}$ for the base space and if $e(X) = 12k$ then $\widetilde{B} \cdot \widetilde{B} = -k$.*

(4-12) was first proved in the case $g = 0$ by Kas and Moishezon [**Ka, M1**]. For general g, [**Man4**] gave the same result.[2] We should note that it is not obvious

---
[2] This is based on Seiler's result. See also [**FM4**].

whether the Lefschetz fibrations are diffeomorphic to complex surfaces in general, and [**Mt2**] proved the result for such Lefschetz fibrations. It is known that the elliptic surface in (4-12) has a cross section $\widetilde{B}$ which is embedded complex analytically if it is equipped with an appropriate complex structure. The claim for $\widetilde{B} \cdot \widetilde{B}$ comes from the adjunction formula. On the other hand, using $\pi_1 X(s_1, s_2, \ldots, s_1, s_2) = 1$, we deduce

(4-14). *For a Lefschetz fibration X with $e(X) > 0$ we have*

$$\pi_1 X \cong \left\{ \alpha_1, \beta_1, \ldots, \alpha_g, \beta_g, q_1, \ldots, q_s \mid \right.$$

$$\left. \prod [\alpha_i, \beta_i] \prod q_j = 1, q_1^{m_1} = \cdots = q_s^{m_s} = 1 \right\}$$

($[\alpha, \beta] = \alpha\beta\alpha^{-1}\beta^{-1}$). *Here we use the same notation for the loops as in §1.*

This group is the fundamental group $\pi_1^{\mathrm{orb}}(\Sigma_g(m_1, \ldots, m_s))$ of the orbifold $\Sigma_g(m_1, \ldots, m_s)$ where B is considered as the orbifold whose singular points of multiplicity m_i are the images of the multiple tori of the same multiplicity. Then if $B = S^2(m, n)$ ($m, n \geq 1$) this group is cyclic of order $\gcd(m, n)$. For the other cases, from (3-1) we deduce

(4-15), [**U2**]. *The diffeomorphism type of any elliptic surface X with $e(X) > 0$ is determined by $e(X)$ and its fundamental group $\pi_1 X$ if $\pi_1 X$ is noncyclic (or equivalently, if the genus of the base space is greater than or equal to 1 or the number of the multiple fibers is greater than or equal to 3). In particular, any such elliptic surfaces are diffeomorphic if they are homeomorphic.*

Let us add the following remark.

(4-16). *Given $g \geq 0$, $s \geq 0$, $(m_1, \ldots, m_s)$ for $m_i \geq 2$, $k \geq 1$ there exists an elliptic surface $\pi : X \to B$ such that $e(X) = 12k$, the genus of B is g, and X contains exactly s multiple fibers each of which has multiplicity m_i.*

There are several ways to construct such elliptic surfaces (see [**Man4**] for example). Let us give one method here. Start with the orbifold $\Sigma_g(\overbrace{2, \ldots, 2}^{2k})$ equipped with the geometric structure (covered by either **H**, **C**, or **CP**^{1}). Then the covering translation ι of order two of the double cover $\Sigma \cong \Sigma_{k-1+2g}$ is complex analytic with respect to some complex structure on Σ. The same claim holds for the double cover $T \cong T^2$ of $S^2(2, 2, 2, 2)$ and its covering translation ι'. Thus the projection to the first factor $\Sigma \times T \to \Sigma$ induces the map $p : \Sigma \times T/\iota \times \iota' \to \Sigma/\iota$ such that each of the fibers over the $2k$ branch points on Σ/ι has just four rational double points. Desingularization of these points by -2 curves yields an elliptic surface $p : X \to B$. This one contains $2k$ fibers of type I_0^*, $e(X) = 12k$, and the genus of B is g. Finally, performing logarithmic transformations on it with given multiplicities we get the desired elliptic surface. As a by-product of these arguments we obtain

(4-17). *Every singular torus fibration over a closed surface with positive Euler number is diffeomorphic to an elliptic surface.*

REMARK (4-18). If the singular fibers F_i ($i = 1, \ldots, t$) that appeared in Table 1 have the monodromy representations P_i with respect to some basis satisfying $\prod P_i = I$, there exists a singular torus fibration $\pi : X \to \Sigma_g$ ($g \geq 0$) that contains F_i as its singular fibers. If $e(X) > 0$, then it is diffeomorphic to an elliptic surface, but π is not necessarily complex analytic. This is because in the complex analytic category we must take account of a J-invariant defined by the period of the general fiber and it restricts the possible choices of the singular fibers that appeared in X. But we do not discuss it any further.

The examples in the next section will show that the above claims ((4-8), (4-10), (4-16)) do not hold in the case $e(X) = 0$. The cases with $\pi_1 X$ cyclic will be discussed in §6.

§5. Topology of elliptic surfaces with Euler number 0

The topology of elliptic surfaces with Euler number 0 is rather well known [**K1,2, Kt, Su1,2 ,I1,2, W3,4, BM**]. But we note that a singular torus fibration $\pi : X \to B$ with $e(X) = 0$ does not necessarily have a complex structure contrary to the case with $e(X) > 0$. Thus in this section we determine when it admits a complex structure and describe its structure explicitly. If $e(X) = 0$ then every singular fiber in X is a multiple torus. So $\pi : X \to B$ can be considered as a *Seifert fibered 4-manifold* [**CR, Z, Zimm**]. Using the notation in §1 it is determined by the following data:

(1) the monodromy representations $A_i, B_i \in \mathrm{SL}(2, \mathbf{Z})$ of $\bar{\alpha}_i, \bar{\beta}_i$ with respect to the basis (ℓ, h) of the general fiber $(\prod_{i=1}^g [A_i, B_i] = I)$;
(2) the type of the ith multiple torus (m_i, a_i, b_i) with respect to the curves (q_i, ℓ, h) $(\gcd(m_i, a_i, b_i) = 1, m_0 = 1)$.

Here q_i is determined by the lift $\widetilde{B}_0$ of $B_0 = B - \bigcup_{j=0}^s \Delta_j$ as in §1. These data are called the *Seifert invariants of X*:

(∗) $\qquad \{A_1, B_1, \ldots, A_g, B_g, (1, a_0, b_0), (m_1, a_1, b_1), \ldots, (m_s, a_s, b_s)\}.$

As in §4 it is natural to consider the base space B as the orbifold $\Sigma_g(m_1, \ldots, m_s)$. Then $\pi_1 X$ is described as follows:

(5-1).

$$\pi_1 X = \Big\{ \alpha_1, \beta_1, \ldots, \alpha_g, \beta_g, q_0, \ldots, q_s, \ell, h \mid [\ell, h] = [q_i, \ell] = [q_i, h] = 1,$$
$$(\alpha_j \ell \alpha_j^{-1}, \alpha_j h \alpha_j^{-1}) = (\ell, h) A_j, (\beta_j \ell \beta_j^{-1}, \beta_j h \beta_j^{-1}) = (\ell, h) B_j,$$
$$q_0 \ell^{a_0} h^{b_0} = q_j^{m_j} \ell^{a_j} h^{b_j} = 1, \quad \prod_{i=1}^g [\alpha_i, \beta_i] \prod_{j=0}^s q_j = 1 \Big\}.$$

The Seifert invariants depend on the choices of (ℓ, h), $\widetilde{B}_0$, or the pull-backs of the fibration by automorphisms of B. For example, by replacing $\widetilde{B}_0$ by a new one we can assume that $a_0 = b_0 = 0$ if $t \geq 1$, but here we do not normalize the invariants. Instead, we assume that any two fibrations that are diffeomorphic by fiber-preserving diffeomorphisms are identified. (If $a_0 = b_0 = 0$, then we can remove $(1, a_0, b_0)$ from the above representation.) On the other hand, for an elliptic surface $\pi : X \to B$ with $e(X) = 0$ the period of its general fiber is constant and the monodromies are the automorphisms of the elliptic curves. So they are the compositions of some periodic matrices and translations along the elliptic curves themselves. But we note that since the fibrations of Seifert fibered 4-manifolds are not unique in general, we need a little more argument to determine when they are diffeomorphic to elliptic surfaces.

We will discuss this problem according as the base orbifolds are (A) hyperbolic, (B) euclidean, (C) spherical or bad, respectively (see §3).

REMARK (5-2). If all monodromies in (∗) are trivial ($A_i = B_i = I$), then (∗) corresponds to the following elliptic surface in Kodaira's notation [**K2**]:
$$L_{x_1}\left(m_1, \frac{a_1 + b_1\omega}{m_1}\right) \cdots L_{x_s}\left(m_s, \frac{a_s + b_s\omega}{m_s}\right)(B \times E).$$

Here B is a Riemann surface of genus g and E is a Riemann surface of genus 1 with period ω. The notation L_{x_i} means that the ith logarithmic transformation is performed over $x_i \in B$.

DEFINITION (5-3). If the monodromies of the Seifert fibered 4-manifold $\pi : X \to B$ are all trivial, then with respect to the representation (∗) above we define
$$\varepsilon(X) = \left(\sum_{i=0}^{s}\frac{a_i}{m_i}, \sum_{i=0}^{s}\frac{b_i}{m_i}\right) \in \mathbf{Q}^2 \mod \mathrm{GL}(2, \mathbf{Z}).$$

(Here $m_0 = 1$.) This is well defined up to the action of $\mathrm{GL}(2, \mathbf{Z})$ since it depends on the choice of (ℓ, h).

(5-4). *Any two Seifert fibered 4-manifolds whose base orbifolds belong to different classes in* (A), (B), *and* (C) *above have distinct fundamental groups.*

(A) The case where B is hyperbolic

In this case the fibration for $\pi : X \to B$ is unique in the following sense.

(5-5), [**Z**]. *Any two Seifert fibered 4-manifolds over hyperbolic base orbifolds with the same fundamental group are diffeomorphic by a fiber-preserving diffeomorphism.*

On the other hand, in this case X does not necessarily have a complex structure.

(5-6), [**U4**]. *The above X has a complex structure if and only if all the monodromies of X are powers of a common periodic matrix. The set of all such X coincides with the class of all complex surfaces of $\mathrm{kod} X = 1$ with $c_2(X) = 0$ (which are necessarily elliptic surfaces).*

(5-7), [**U5**]. *After the pull-back by some automorphism of the base B such X has a fibration whose monodromy representations satisfy $A_i = B_i = I$ ($i \geq 2$),*
$$B_1 = I, \quad A_1 = \begin{pmatrix} 1 & 1 \\ -1 & 0 \end{pmatrix}, \begin{pmatrix} 0 & 1 \\ -1 & -1, \end{pmatrix}, \begin{pmatrix} 0 & 1 \\ -1 & 0 \end{pmatrix}, -I, \text{ or } I.$$

(B) The case when B is euclidean

(5-8), [**U3**]. *The diffeomorphism type of any Seifert fibered 4-manifold over a euclidean base orbifold is determined by its fundamental group. But its fibering structure is not unique in general.*

If the genus of the base B is 0, then X is diffeomorphic to an elliptic surface, but if $B = T^2$, then it does not necessarily have a complex structure (see [**SF**] for the classification of T^2-bundles over T^2). In the following list we give the Seifert invariants for those that admit complex structures.

(B-1) $\{I, I, (1, 0, 0)\} \cong T^4$ (the complex torus of dimension two).

(B-2) Seven hyperelliptic surfaces [Su2, BM]. Each of them has two fibrations. These cover all Seifert fibered 4-manifolds over the euclidean base orbifolds of genus 0 with $\varepsilon(X) = (0,0)$:

$$\{-I, I, (1, -1, 0)\} \cong \{(2, -1, 0), (2, 1, 0), (2, 0, -1), (2, 0, 1)\},$$

$$\{-I, I, (1, 0, 0)\} \cong \{(2, 1, 0), (2, 1, 0), (2, -1, 0), (2, -1, 0)\},$$

$$\left\{\begin{pmatrix} 0 & -1 \\ 1 & -1 \end{pmatrix}, I, (1, -1, 0)\right\} \cong \{(3, -1, 0), (3, 0, -1), (3, 1, 1)\},$$

$$\left\{\begin{pmatrix} 0 & -1 \\ 1 & -1 \end{pmatrix}, I, (1, 0, 0)\right\} \cong \{(1, -1, 0), (3, 1, 0), (3, 1, 0), (3, 1, 0)\},$$

$$\left\{\begin{pmatrix} 0 & -1 \\ 1 & 0 \end{pmatrix}, I, (1, -1, 0)\right\} \cong \{(2, -1, 0), (4, 1, 1), (4, 1, -1)\},$$

$$\left\{\begin{pmatrix} 0 & -1 \\ 1 & 0 \end{pmatrix}, I, (1, 0, 0)\right\} \cong \{(1, -1, 0), (2, 1, 0), (4, 1, 0), (4, 1, 0)\},$$

$$\left\{\begin{pmatrix} 1 & -1 \\ 1 & 0 \end{pmatrix}, I, (1, 0, 0)\right\} \cong \{(1, -1, 0), (2, 1, 0), (3, 1, 0), (6, 1, 0)\}.$$

(B-3) Primary Kodaira surfaces $\{I, I, (1, a, 0)\}$ $(a \neq 0)$.
Every T^2-bundle over T^2 with $b_1 = 3$ is of the form

$$\left\{\begin{pmatrix} 1 & \lambda \\ 0 & 1 \end{pmatrix}, I, (1, a, 0)\right\} \quad ((\lambda, a) \neq (0, 0))$$

[SF], each of which is diffeomorphic to the one of the form in (B-3) (but the bundle structure is different). The complex analytic fibrations only correspond to those of the form in (B-3).

(B-4) Secondary Kodaira surfaces.
These coincide with Seifert fibered 4-manifolds over euclidean base orbifolds of genus 0 with $\varepsilon(X) \neq (0, 0)$.

(C) The case where B is spherical or "bad"

In this case X is always diffeomorphic to an elliptic surface (see the remark in §2 and [Su1, Kt]).

(C-1) Ruled surfaces of genus 1 [Su1].
$$\{(n, a, b), (n, -a, -b)\} \cong S^2 \times T^2 \quad (\gcd(n, a, b) = 1, n \geq 1),$$
$$\{(2, 0, 1), (2, -1, 0), (2, 1, 1)\} \quad \text{the } S^2\text{-bundle over } T^2 \text{ with } w_2 \neq 0.$$

(C-2) Hopf surfaces.
Every Hopf surface is a bundle over S^1 with fiber a spherical 3-manifold F whose monodromy t has period 1, 2, or 3 [Kt]. We will denote it by $F \times_t S^1$. Such a bundle is a Hopf surface if and only if it appears in the following list (we use the same notation as in §3):

(i) $L(p, q) \times S^1 \cong \{(n_1, a_1, b_1), (n_2, a_2, b_2)\}$ where $\varepsilon = (\sum \frac{a_i}{n_i}, \sum \frac{b_i}{n_i}) \neq (0, 0)$. The relation between (p, q) and (n_i, a_i, b_i) is given by

$$k_i = \gcd(a_i, b_i),$$
(*) $\quad p = \gcd(n_2 k_1 + n_1(s_1 a_2 + t_1 b_2), (-a_2 b_1 + a_1 b_2)/k_1),$
$$q = -u_1 n_2 + v_1(s_1 a_2 + t_1 b_2)$$

where s_i, t_i, u_i, v_i are defined by $a_i s_i + b_i t_i = k_i$, $n_i u_i + k_i v_i = 1$.

REMARK. If we assume $L(-p,q) = L(p,-q) = -L(p,q)$ then $L(-p,q) \times S^1 \cong L(p,q) \times S^1$ and so the signs of p and q are not essential. In the representation of $(*)$ there is some ambiguity for the sign of k_i since a_i and b_i may be nonpositive. But if k_i is replaced by $-k_i$, then we can replace (s_i, t_i) by $(-s_i, -t_i)$ and (u_i, v_i) by $(u_i, -v_i)$ to get the well-determined p and q up to sign.

(ii) $L(p,q) \times_\sigma S^1 \cong \{(2,1,0), (2,-1,b), (\alpha,0,\beta)\}$ (the relation between p, q, α, β, b is given by §3, (L)).
(iii) $F_1 \times S^1 \cong \{(1,b,0), (2,1,0), (2,1,0), (n,k,0)\}$ $(n \geq 2)$.
(iv) $F_1 \times_\sigma S^1 \cong \{(2n,n,k), (2,-1,b), (2,0,1)\}$.
(v) $F_1' \times_\rho S^1 \cong \{(2,0,1), (3,1,0), (3,-1,b)\}$.
(vi) $F_2 \times S^1 \cong \{(1,b,0), (2,1,0), (3,1,0), (3,k,0)\}$.
(vii) $F_2 \times_\sigma S^1 \cong \{(2,1,b), (4,-2,1), (3,0,1)\}$.
(viii) $F_3 \times S^1 \cong \{(1,b,0), (2,1,0), (3,1,0), (4,k,0)\}$.
(ix) $F_4 \times S^1 \cong \{(1,b,0), (2,1,0), (3,1,0), (5,k,0)\}$.

Finally taking account of the other Seifert fibrations admitting no complex structures which are not discussed here we can see the following.

(5-9), [U4]. *Any two Seifert fibered 4-manifolds are diffeomorphic if they are homeomorphic.*

§6. Topology of elliptic surfaces with positive Euler numbers (II) —the cases with cyclic fundamental groups

The preceding results show that an elliptic surface $\pi : X \to B$ has a cyclic fundamental group if and only if the genus of B is 0, the number of multiple fibers in X is less than or equal to two, and $e(X) > 0$ (which is a multiple of 12). Furthermore, its diffeomorphism type is determined by $e(X)$ ($= 12k$), the multiplicities (m_1, m_2) of the multiple fibers, and so we denote it by $S_k(m_1, m_2)$ (which was already used in §2). We assume that $m_i \geq 1$ and then this representation makes sense even if the number of multiple fibers is less than two. In this case $\pi_1 S_k(m_1, m_2)$ is cyclic of order $\gcd(m_1, m_2)$. Contrary to the other cases, the homeomorphism classification for these classes is completely different from the diffeomorphism classification of them. We can say nothing about them without Freedman and Donaldson's theories. So we will discuss them in the following two subsections.

(I) The classification of the homeomorphism types

First let us recall the results of Freedman and Hambleton-Kreck about the homeomorphism classification of a closed oriented smooth 4-manifold X with finite cyclic fundamental group.

(6-1), [F, FQ]. *If $\pi_1 X = 1$, then the homeomorphism type of X is determined by q_X.*

(6-2), [HK1]. *If $\pi_1 X = \mathbf{Z}_r$ for r odd, then the homeomorphism type of X is determined by the intersection form q_X on $H_2(X, \mathbf{Z})/$Torsion.*

(6-3), [HK2, HK4]. *If $\pi_1 X = \mathbf{Z}_r$ for r even, then the homeomorphism type of X is determined by the intersection form q_X over $H_2(X, \mathbf{Z})/$Torsion and the w_2-type of X. Here X is said to be of w_2-type* I, II, *or* III *according as* (I) $w_2(\widetilde{X}) \neq 0$, (II) $w_2(X) = 0$, *or* (III) $w_2(X) \neq 0$, $w_2(\widetilde{X}) = 0$, *respectively, where $\widetilde{X}$ is the universal covering of X.*

REMARK. The Kirby-Siebenmann obstruction is not necessary for our cases (it is always zero) since we only consider the smooth manifolds.

Let us add one supplementary remark.

(6-4), THE CONSTRUCTION OF L_r. Take a neighborhood B^3 of one point in a three-dimensional lens space $L(r,s)$ and let L_r be a manifold obtained from $L(r,s) \times S^1$ by removing $B^3 \times S^1$ and reattaching $S^2 \times B^2$. The definition of this manifold does not depend on the choice of s nor on the way of attaching $S^2 \times B^2$. L_r satisfies $\pi_1 L_r \cong \mathbf{Z}_r$ and $H_*(L_r, \mathbf{Q}) \cong H_*(S^4, \mathbf{Q})$. The universal cover of L_r is $(r-1)S^2 \times S^2$ (if $r = 1$ then $L_1 \cong S^4$). □

The same L_r was described as a 4-manifold with T^2-action [**OR, P**], and also as a torus fibration in the sense of [**Mt1, Iw**]. If r is odd, then any **Q**-homology 4-sphere whose fundamental group is $\mathbf{Z}_r$ is homeomorphic to L_r according to (6-2) (this is not true if r is even [**P, Iw, HK4**]).

Let us apply the above theorems to $S \cong S_k(p_1, p_2)$ with $\pi : S \to S^2$. Put $r = \gcd(p_1, p_2)$, $p_i' = p_i/r$ and fix the integers a_1, a_2 satisfying $a_1 p_1' + a_2 p_2' = 1$. Recall that $S_k(p_1, p_2)$ is obtained from $S_k = S_k(1,1)$ by removing the neighborhoods of two general fibers and attaching the neighborhoods of multiple tori F_1 and F_2 of multiplicities p_1 and p_2 respectively. Take the basis (ℓ, h) of the general fiber and the curves q_1, q_2 as in §1 and attach $N(F_i)$ as in (4-10). Then the 2-cycles

$$f = a_2(q_1 \times \ell) + a_1(q_2 \times \ell), \quad c = p_1'(q_1 \times \ell) - p_2'(q_2 \times \ell)$$

satisfy

(∗) $\quad f \cdot f = 0, \quad f \cdot g = 1, \quad g \cdot g = -r(p_1' p_2')^2 k - (p_1' + p_2')^2, \quad rc = 0,$

where g is an appropriate 2-cycle.

Then we have

(6-5) $\quad H_2(S, \mathbf{Z}) \cong \mathbf{Z}^{12k-2} \oplus \mathbf{Z}_r, \quad q_s \cong kE_8 \oplus (2k-2)U \oplus U'.$

Here U' is generated by f and g, and

$$U' \cong \begin{cases} U & rk \equiv 0 \pmod 2, \; p_1' \equiv p_2' \equiv 1 \pmod 2, \\ (1) \oplus (-1) & \text{otherwise.} \end{cases}$$

The $\mathbf{Z}_r$-factor is generated by c. In particular, $e(S) = 12k$ and $\sigma(S) = -8k$. On the other hand, the canonical divisor K_S of S is given by

$K_S = (k-2)F + (p_1-1)F_1 + (p_2-1)F_2 = (rp_1' p_2' k - p_1' - p_2')f + (a_2 - a_1)c.$

Furthermore note that

(6-6). *The universal covering of $S = S_k(p_1, p_2)$ is $S_{rk}(p_1', p_2')$.*

These facts determine q_S and also the w_2-type of S if r is even. Hence the homeomorphism classification of S is given as follows.

(6-7). *$S_k(p_1, p_2)$ is homeomorphic to $S_{\tilde{k}}(\tilde{p}_1, \tilde{p}_2)$ if and only if $k = \tilde{k}$, $\gcd(p_1, p_2) = \gcd(\tilde{p}_1, \tilde{p}_2)$ (which we denote by r) and one of the following conditions is satisfied:*
(1) $r \equiv 1 \pmod 2$, $k \equiv 0 \pmod 2$, and $\frac{p_1+p_2}{r} \equiv \frac{\tilde{p}_1+\tilde{p}_2}{r} \pmod 2$,
(2) $r \equiv 1 \pmod 2$ and $k \equiv 1 \pmod 2$,

(3) $r \equiv 0$ (mod 2) and $\frac{p_1+p_2}{r} \equiv \frac{\tilde{p}_1+\tilde{p}_2}{r}$ (mod 2).

In particular, if $r \equiv 1$ (mod 2) we have

(6-8). $S_k(p_1, p_2)$ is homeomorphic to $L_r \sharp \frac{k}{2}(K3) \sharp (\frac{k}{2}-1) S^2 \times S^2$ if $k \equiv 0$ (mod 2) and $\frac{p_1}{r} \equiv \frac{p_2}{r} \equiv 1$ (mod 2) and is homeomorphic to $L_r \sharp (2k-1) \mathbf{CP}^2 \sharp (10k-1) \overline{\mathbf{CP}}^2$ otherwise.

(II) On the diffeomorphism types

Let us start with the following conjecture.

(6-9). If $S_k(p_1, p_2) \cong S_k(p'_1, p'_2)$ $(p_1 \geq p_2, p'_1 \geq p'_2)$, then $k = k'$, $p_1 = p'_1$, and $p_2 = p'_2$ except for the following cases:

$$S_1(p, 1) \cong S_1(1, 1) \cong \mathbf{CP}^2 \sharp 9\overline{\mathbf{CP}}^2.$$

This conjecture is not yet completely settled, but according to many deep studies it is quite likely that we are close to the complete solution [**D5, FM1, FM2, OV1, OV2, Mon, O, Mai, LO, Ko3**]. At present the strongest results are due to Friedman and Morgan and so we will state their results in the simply connected cases.[3]

(6-10), [**FM1**]. *Every Dolgachev surface $S_1(p, q)$ $(p, q \geq 2, \gcd(p, q) = 1)$ is homeomorphic but not diffeomorphic to the rational surface $S_1(1, 1)$. The class of Dolgachev surfaces contains infinitely many examples which are not mutually diffeomorphic. In particular, if $p \neq q$ then we have $S_1(2, p) \not\cong S_1(2, q)$.*

(6-11), [**FM2**]. *Let $S_k(p, q)$ and $S_k(p', q')$ be simply connected elliptic surfaces with $k \geq 2$ $(p, q, p', q' \geq 1)$. If $pq \neq p'q'$, then $S_k(p, q) \not\cong S_k(p', q')$.*

The last theorem implies that any homotopy $K3$ surface with at least one multiple fiber is not diffeomorphic to $K3$. Friedman and Morgan also proved the following.

(6-12). *If $\widetilde{S} = S_k(p, q) \sharp r\overline{\mathbf{CP}}^2$ $(\gcd(p, q) = 1)$ is diffeomorphic to a connected sum of two smooth 4-manifolds A and B such that q_B is negative definite then $H_2(B)$ is contained in the $H_2(r\overline{\mathbf{CP}}^2)$-factor.*

According to Donaldson's theorems, if $\widetilde{S} \cong A \sharp B$ for $k \geq 2$ $(b_2^+(\widetilde{S}) \geq 3)$, then either q_A or q_B is negative definite [**D4**]. Furthermore, if q_B is negative definite then it is a direct sum of (-1) [**D1, D3**]. These results are derived from the computation of Donaldson invariants defined via the moduli spaces of SU(2) (or SO(3)) instantons on the base 4-manifolds. (6-11) was obtained by some algebro-geometric computation based on the results in [**Fr, FMM**]. For the details of these results see the book written by Friedman and Morgan [**FM4**]. Now we state some extra conclusions.

(6-13), [**FM1, FM2**]. *For a natural number r, if $p \neq q$ then $S_1(2, p) \sharp r\overline{\mathbf{CP}^2} \not\cong S_1(2, q) \sharp r\overline{\mathbf{CP}^2}$. If $pq \neq p'q'$ then $S_k(p, q) \sharp r\overline{\mathbf{CP}^2} \not\cong S_k(p', q') \sharp r\overline{\mathbf{CP}^2}$ $(k \geq 2)$.*

(6-14). *If $r = \gcd(p, q) \equiv 1$ (mod 2), then $S_k(p, q)$ is homeomorphic but not diffeomorphic to a connected sum of L_r and copies of $K3$, $S^2 \times S^2$ (if $w_2 = 0$) or a connected sum of L_r and copies of $\pm \mathbf{CP}^2$ (if $w_2 \neq 0$) except for the rational surface*

---

[3] See also [**Ba, EO1, EO2**]. The diffeomorphism classification of elliptic surfaces with $p_g = 1$ was completed by Morgan-O'Grady and Bauer.

and $K3$. *In fact, according to Donaldson's theorem* **[D4]** *these connected sums are not diffeomorphic to complex surfaces.*

REMARK (6-15). No simply connected elliptic surface S is of the form $S \cong S' \sharp \overline{\mathbf{CP}^2}$ except for the rational surface ((6-12)). Likewise S is not of the form $S' \sharp \mathbf{CP}^2$ nor of the form $S' \sharp S^2 \times S^2$. This follows from (6-12) and the results in **[D4, DK]**.

REMARK (6-16). The computation of the Donaldson invariants also provides deep results on the algebraic surfaces of general type (**[Ko1, Ko2, Ko3, OV, E, S, M2]**).

Let us state that (6-14) is generalized as follows.

(6-17), **[HK4]**. *Suppose that S is an elliptic surface with $\pi_1 S$ finite and $p_g(S) > 0$. Then there exists at least one smooth manifold that is homeomorphic to S but not diffeomorphic to any complex surface. In the case where $p_g(S) = 0$ the same conclusion holds if S is replaced by $S \sharp \overline{\mathbf{CP}^2}$.*

In fact, **[HK4]** shows the existence of a 4-manifold homeomorphic to S whose universal covering is a connected sum of two smooth manifolds with indefinite intersection forms.

§7. Decomposition of elliptic surfaces and related topics

The results in §6 show that there are tight restrictions on smooth connected sum decompositions for elliptic surfaces. So let us introduce decompositions of other types here.

(I) Decompositions as fiber sums

DEFINITION (7-1), **[Man2]**. Let V and W be singular torus fibrations, $N(F) \subset V$ a neighborhood of a general fiber F of V, and $\varphi : N(F) \to W$ an orientation-reversing and fiber-preserving embedding. Then

$$V \sharp_\varphi W = (W - \varphi(N(F))) \cup_\varphi (V - N(F))$$

is called the *fiber sum of V and W by φ*.

On the other hand, the following claim was proved for the neighborhood $N(F)$ of the general fiber F of the rational elliptic surface S_1.

(7-2), **[Mat1, G1]**. *Every orientation-preserving self-diffeomorphism of $\partial(S_1 - N(F)) \cong T^3$ extends to that of $S_1 - N(F)$.*

This leads to the following conclusion.

(7-3), **[G4]**. $S_k(p_1, \ldots, p_n)$ *(in the sense given in §2) is a fiber sum of $S_1(p_1, \ldots, p_n)$ and $k - 1$ copies of S_1. This sum does not depend on the gluing maps.*

(7-4), **[Mat1, GM]**. *Let φ be an orientation-reversing diffeomorphism of $\partial(S_1 - N(F)) \cong T^3$ and put $K_\varphi = (S_1 - N(F)) \cup_\varphi (S_1 - N(F))$. Then for any such φ we have $K_\varphi \cong K3$.*

The last result is a starting point for the construction of the examples by Gompf-Mrowka which we will state later. On the other hand, Mandelbaum **[Man2]** gave a certain trick concerning fiber sums which leads to the following results.

(7-5). *Let S and S' be any simply connected elliptic surfaces. Then the following manifolds are diffeomorphic either to connected sums of copies of $\pm\mathbf{CP}^2$ or to connected sums of copies of $K3$ and $S^2 \times S^2$:*
 (1) [M1, Man2] $S \sharp \mathbf{CP}^2$,
 (2) [Man2] $S \sharp S^2 \times S^2$,
 (3) [G4] $S \sharp \overline{S}'$.

Note that these results form sharp contrasts to (6-13).

(II) Decompositions by homology spheres

Freedman and Taylor showed the following (here we describe it in a slightly modified form using the result in [**F**], etc.).

(7-6), [**FT**]. *For any simply connected closed smooth 4-manifold X with $q_X \cong A_1 \oplus \cdots \oplus A_k$ there exist mutually disjoint homology 3-spheres $\Sigma_1, \ldots, \Sigma_{k-1}$ that are smoothly embedded in X such that by cutting X with them we obtain*

$$X \cong X_1 \cup_{\Sigma_1} X_2 \cup \cdots \cup_{\Sigma_{k-1}} X_k \quad \text{with } q_{X_i} \cong A_i.$$

Such decompositions are not easy to obtain in general, but in the case of simply connected elliptic surfaces $S_k(p,q)$ we have

(7-7), [**G1, U6**]. *Corresponding to the direct sum decomposition of the intersection form in (6-5), $S_k(p,q)$ has the following decompositions by some Brieskorn homology spheres. (Each of them does not depend on the gluing maps on the boundaries of the pieces. If $k = 1$ then $Q = \emptyset$, $Q' = B^4$ below.)*
 (1) $S_k(p,q) \cong N_k(p,q) \cup Q \cup (kE_8)$,
 (2) $S_k(p,q) \cong N_k(p,q) \cup E_8' \cup Q' \cup (k-1)E_8$
where

$$q_{N_k(p,q)} \cong U', \quad q_{E_8} \cong q_{E_8'} \cong E_8, \quad q_Q \cong q_Q' \cong (2k-2)U,$$

$$\partial N_k(p,q) \cong -\Sigma(2,3,6k-1), \quad \partial E_8 \cong \Sigma(2,3,5),$$

$$\partial E_8' \cong -\Sigma(2,3,6k-5) \cup \Sigma(2,3,6k-1),$$

$$\partial Q \cong \Sigma(2,3,6k-1) \cup k(-\Sigma(2,3,5)),$$

$$\partial Q' \cong \Sigma(2,3,6k-5) \cup (k-1)(-\Sigma(2,3,5)).$$

The definition of each submanifold above is as follows. First, E_8 is an E_8-plumbing corresponding to the graph shown in Figure 6 or the framed link. $N_k(1,1)$ is defined as a regular neighborhood of the sum of the fiber of type II and the cross section $\tilde{B}$ of $S_k(1,1)$ with respect to the fibration of $S_k(1,1)$ that contains just k fibers of type II and k fibers of type II* (see (4-18)). $N_k(p,q)$ is obtained from $N_k(1,1)$ by performing logarithmic transformations of multiplicity p and q on two general fibers (by the same reason stated in §4 the diffeomorphism type of $N_k(p,q)$ depends only on k, p, q). In [**G1**] $N_k(p,q)$ is called a nucleus. E_8' is obtained from a neighborhood of the sum of the fiber of type II* and $\tilde{B}$ by removing $N_k(1,1)$ contained in it (cf. (4-6)). Q and Q' are defined as the remaining parts respectively. According to the result in [**G1**] and a similar observation we have

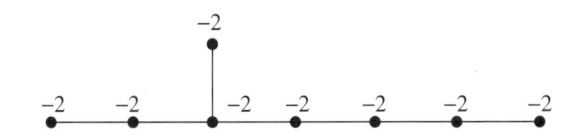

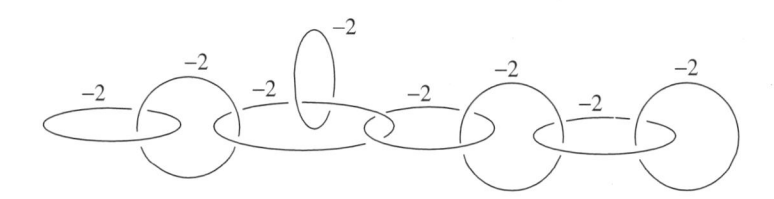

FIGURE 6

(7-8), [**G1**]. $Q \cup (kE_8)$ in (7-7) is diffeomorphic to the Milnor fiber of $\Sigma(2, 3, 6k - 1)$. Likewise $Q' \cup (k - 1)E_8$ is diffeomorphic to the Milnor fiber of $\Sigma(2, 3, 6k - 5)$.[4]

Finally let us state the examples given by Gompf-Mrowka. Take a map $\varphi : \partial(S_1 - N(F)) \to \partial(S_1 - N(F))$ in (7-4) so that it induces a cyclic permutation of three S^1-factors of $\partial(S_1 - N(F)) \cong S^1 \times S^1 \times S^1$ ($* \times S^1 \times S^1$ corresponds to the general fiber). Then $K_\varphi \cong K3$ although we cannot define the natural fibration of K_φ. On the other hand, by the definition of φ we can see that three mutually disjoint copies of $N_2(1, 1)$ above are contained in K_φ [**GM**]. Then replacing each copy by $N_2(p_i, q_i)$ ($i = 1, 2, 3$) as before we get the manifold $K(p_1, q_1, p_2, q_2, p_3, q_3)$. It is homeomorphic to $K3$ if

(∗) $\qquad \gcd(p_i, q_i) = 1, \quad p_i \equiv q_i \equiv 1 \pmod 2 \quad$ for any i.

[**GM**] shows the following.

(7-9), [**GM**]. If $K(p_1, q_1, p_2, q_2, p_3, q_3) \cong K(p'_1, q'_1, p'_2, q'_2, p'_3, q'_3)$ (and all p_i, q_i, p'_i, q'_i satisfy (∗)) then the set $\{p_1q_1, p_2q_2, p_3q_3\}$ is the same as $\{p'_1q'_1, p'_2q'_2, p'_3q'_3\}$ up to permutation.

(7-10), [**GM**]. Any $K(p_1, q_1, p_2, q_2, p_3, q_3)$ above is not diffeomorphic to a connected sum of smooth 4-manifolds M and N with $H_2(M) \neq 0$ and $H_2(N) \neq 0$.

These classes are not diffeomorphic to complex surfaces except for $K(p_1, q_1, 1, 1, 1, 1) \cong K(1, 1, p_1q_1, 1, 1) \cong K(1, 1, 1, 1, p_1, q_1)$, which is diffeomorphic to $S_2(p_1, q_1)$. They are remarkable examples because they provide the first (infinitely many) counterexamples to the following conjecture (which originated from Thom according to [**FM3**]).

---

[4] In fact, the above decompositions of the elliptic surfaces coincide with those obtained by compactifying the Milnor fibers of the Brieskorn singularities (see [**EO1**]).

(7-11) CONJECTURE (now denied). *Any smooth closed simply connected 4-manifold is diffeomorphic to a connected sum of S^4, the algebraic surfaces and the algebraic surfaces with orientation reversed.*[5]

As special cases of Donaldson invariants described in §6, there are the **Z**-valued invariants defined by the number of the zero-dimensional components of the moduli spaces counted with sign consisting of anti-self-dual connections on certain SO(3)-bundles (they are called simple invariants, [**DK, GM**]). The results in [**GM**] were obtained by formulating how the simple invariants change according to the logarithmic transforms and by estimating these invariants of the prescribed manifolds. Furthermore, Kametani-Sato [**KS**] computed these invariants of S_k (in §6) and gave many examples homeomorphic but not diffeomorphic to S_k ($k \geq 2$) which are obtained from S_k in a way similar to those in [**GM**] derived from S_2 (the $K3$ surface). Moreover, using these results [**GM, KS**] we can get the following.

(7-12), [**U7**]. *Let S be an elliptic surface with $\pi_1 S$ finite and $p_g(S) \geq 1$. (In the case where $\pi_1(S)$ is cyclic, assume further that its order is odd and $p_g(S)$ is even.) Then there exist infinitely many smooth manifolds that are homeomorphic to S but not diffeomorphic to any complex surface.*

The universal coverings of the examples in (7-12) have the same properties as those in (7-10); so they are different from those in (6-17). The above results suggest that the logarithmic transforms, the decomposition by homology spheres, and the associated replacement of some parts of the original manifolds produce some interesting 4-manifolds around elliptic surfaces. But an attempt at further development has just begun.

References

[BPV] W. Barth, C. Peters, and A. Van de Ven, *Compact complex surfaces*, Ergeb. Math. Grenzgeb. (3), Springer-Verlag, Berlin and New York, 1984.

[B] F. A. Bogomolov, *Classification of surfaces of class VII with $b_2 = 0$*, Math. USSR-Izv. **10** (1976), 255–269; *Surfaces of class VII_0 and affine geometry*, Math. USSR-Izv. **21** (1983), 31–73.

[BM] E. Bombieri and D. Mumford, *Enriques' classification in characteristic p*. II, Complex Analysis and Algebraic Geometry, Iwanami, Tokyo, 1977, pp. 23–44.

[CR] P. E. Conner and F. Raymond, *Holomorphic Seifert fiberings*, Proc. Second Conf. on Compact Transformation Groups, Lecture Notes in Math., vol. 299, Springer-Verlag, Berlin and New York, 1972, pp. 124–204.

[D1] S. K. Donaldson, *An application of gauge theory to four dimensional topology*, J. Differential Geom. **18** (1983), 279–315.

[D2] ——, *Connections, cohomology and the intersection forms of four manifolds*, J. Differential Geom. **24** (1986), 275–341.

[D3] ——, *The orientation of Yang-Mills moduli spaces and 4-manifold topology*, J. Differential Geom. **26** (1987), 397–428.

[D4] ——, *Polynomial invariants for smooth four manifolds*, Topology **29** (1990), 257–315.

[D5] ——, *Irrationality and the h-cobordism conjecture*, J. Differential Geom. **26** (1987), 141–168.

[DK] S. K. Donaldson and P. B. Kronheimer, *The geometry of four manifolds*, Oxford Math. Monographs, Oxford Univ. Press, New York, 1990.

[Do] I. Dolgachev, *Algebraic surfaces with $p_g = q = 0$*, Algebraic Surfaces, Liguori, Napoli, 1981.

[E] W. Ebeling, *An example of two homeomorphic, nondiffeomorphic complete intersection surfaces*, Invent. Math. **99** (1990), 651–654.

---

[5] Now the other counterexamples (e.g., those homeomorphic to some algebraic surfaces of general type) are constructed by Fintushel-Stern and others.

[F]   M. Freedman, *The topology of four-dimensional manifolds*, J. Differential Geom. **17** (1982), 337–453.
[FM1]   R. Friedman and J. Morgan, *On the differentiable classification of algebraic surfaces*. I, II, J. Differential Geom. **77** (1988), 297–369, 371–398.
[FM2]   _____, *Complex versus differentiable classification of algebraic surfaces*, Topology Appl. **32** (1989), 135–139.
[FM3]   _____, *Algebraic surfaces and 4-manifolds: Some conjectures and speculations*, Bull. Amer. Math. Soc. (N. S.) **18** (1988), 1–19.
[FMM]   R. Friedman, B. Moishezon, and J. Morgan, *On the C^∞ invariance of the canonical classes of certain algebraic surfaces*, Bull. Amer. Math. Soc. (N. S.) (1987), 283–286.
[FQ]   M. Freedman and F. Quinn, *Topology of 4-manifolds*, Princeton Math. Ser., vol. 39, Princeton Univ. Press, Princeton, NJ, 1990.
[FT]   M. Freedman and L. Taylor, *Λ-splitting 4-manifolds*, Topology **16** (1977), 181–184.
[Fr]   R. Friedman, *Rank 2 vector bundles over regular elliptic surfaces*, Invent. Math. **96** (1989), 283–332.
[G1]   R. Gompf, *Nuclei of elliptic surfaces*, Topology **30** (1991), 479–512.
[G2]   _____, *Stable diffeomorphism of compact 4-manifolds*, Topology Appl. **18** (1984), 115–120.
[G3]   _____, *On sums of algebraic surfaces*, Invent. Math. **94** (1988), 171–174.
[G4]   _____, *Sums of elliptic surfaces*, J. Differential Geom. **34** (1991), 93–114.
[GM]   R. Gompf and T. S. Mrowka, *Irreducible four manifolds need not be complex*, Ann. Math. **138** (1993), 61–111.
[HK1]   I. Hambleton and M. Kreck, *On the classification of topological 4-manifolds with finite fundamental group*, Math. Ann. **280** (1988), 85–104.
[HK2]   _____, *Smooth structures on algebraic surfaces with cyclic fundamental group*, Invent. Math. **91** (1988), 53–59.
[HK3]   _____, *Smooth structures on algebraic surfaces with finite fundamental group*, Invent. Math. **102** (1990), 109–114.
[HK4]   _____, *Cancellation, elliptic surfaces and the topology of certain four-manifolds*, J. Reine Angew. Math. **444** (1993), 79–100.
[HKK]   J. Harer, A. Kas, and R. Kirby, *Handlebody decompositions of complex surfaces*, Mem. Amer. Math. Soc., No. 82 (1986).
[I1]   S. Iitaka, *Deformations of compact complex surfaces*, Global Analysis: Papers in Honor of Kodaira, Univ. of Tokyo Press, and Princeton Univ. Press, Princeton, NJ, 1969, pp. 267–272.
[I2]   _____, *Deformations of compact complex surfaces*. II, J. Math. Soc. Japan **22** (1970), 247–261; III, J. Math. Soc. Japan **23** (1971), 692–704.
[In]   M. Inoue, *On surfaces of class VII_0*, Invent. Math. **24** (1974), 269–310.
[Iw]   Z. Iwase, *Good torus fibrations with twin singular fibers*, Japan J. Math. (N. S.) **10** (1984), 321–352.
[K1]   K. Kodaira, *On compact complex analytic surfaces*. I, Ann. of Math. (2) **71** (1960), 111–152; II, Ann. of Math. (2) **77** (1963), 563–626; III, Ann. of Math. (2) **78** (1963), 1–40.
[K2]   _____, *On the structure of compact complex analytic surfaces*. I, Amer. J. Math. **86** (1964), 751–798; II, Amer. J. Math. **88** (1966), 682–721; III, Amer. J. Math. **90** (1969), 55–83; IV, Amer. J. Math. **90** (1969), 1048–1066.
[K3]   _____, *On homotopy $K3$-surfaces*, Essays on Topology and Related Topics, Springer, Heidelberg, 1970, pp. 58–69.
[Ka]   A. Kas, *On the deformation types of regular elliptic surfaces*, Complex Analysis and Algebraic Geometry, Cambridge Univ. Press, Cambridge, 1977, pp. 107–111.
[Kt]   Ma. Kato, *Topology of Hopf surfaces*, J. Math. Soc. Japan **27** (1975), 222–238; *Erratum to "Topology of Hopf Surfaces"*, J. Math. Soc. Japan **41** (1989), 173–174.
[Ki]   R. Kirby, *A calculus for framed links in S^3*, Invent. Math. **45** (1978), 35–56.
[Ko1]   D. Kotschick, *On manifolds homeomorphic to $CP^2 \sharp 8\overline{CP^2}$*, Invent. Math. **95** (1989), 591–600.
[Ko2]   _____, *The topology of algebraic surfaces with $q = p_g = 0$*, Geometry of Low-Dimensional Manifolds 1, vol. 150, London Math. Soc., pp. 55–62.
[Ko3]   _____, *$SO(3)$-invariants for 4-manifolds with $b_2^+ = 1$*, Proc. London Math. Soc. (3) **63** (1991), 426–448.
[Ko4]   _____, *Remarks on geometric structures on compact complex surfaces*, Topology **31** (1992), 317–321.
[KS]   Y. Kametani and Y. Sato, *0-dimensional moduli spaces of stable rank 2 bundles and differentiable structures on regular elliptic surfaces*, preprint, 1991.

[LO] M. Lübke and C. Okonek, *Differentiable structures on elliptic surfaces with finite fundamental group*, Compositio Math. **63** (1987), 217–222.

[LYZ] J. Li, S. T. Yau, and F. Zheng, *A simple proof of Bogomolov's theorem on class VII_0 surfaces with $b_2 = 0$*, Illinois J. Math. **34** (1990), 217–220.

[M1] B. Moishezon, *Complex surfaces and connected sums of complex projective planes*, Lecture Notes in Math., vol. 603, Springer-Verlag, Berlin and New York, 1977.

[M2] _____, *Analogs of Lefschetz theorems for linear systems with isolated singularities*, J. Differential Geom. **31** (1990), 47–72.

[Man1] R. Mandelbaum, *Four dimensional topology: An introduction*, Bull. Amer. Math. Soc. (N. S.) **2** (1980), 1–159.

[Man2] _____, *Decomposing analytic surfaces*, Proc. 1977 Georgia Topology Conference, Geometric Topology, 1979, pp. 147–218.

[Man3] _____, *On the topology of elliptic surfaces*, Adv. Math. Suppl. Stud. **5** (1979), 143–166.

[Man4] _____, *Lefschetz fibrations of Riemann surfaces and decompositions of complex elliptic surfaces*, Contemp. Math., No. 44, Amer. Math. Soc., Providence, RI, 1985, pp. 291–310.

[Mt1] Y. Matsumoto, *Good torus fibrations*, Contemp. Math., No. 35, Amer. Math. Soc., Providence, RI, 1984, pp. 375–397.

[Mt2] _____, *Diffeomorphism types of elliptic surfaces*, Topology **25** (1986), 549–564.

[Mt3] _____, *Torus fibrations over the 2-sphere with the simplest singular fibers*, J. Math. Soc. Japan **37** (1985), 605–636.

[Mat1] T. Matumoto, *Extension problem of diffeomorphisms of a 3-torus over some 4-manifolds*, Hiroshima Math. J. **14** (1984), 189–201.

[Mat2] _____, *On diffeomorphisms of a K3 surface*, Algebraic and Topological Theories—to the memory of T. Miyata, Kikokuniya, Tokyo, 1985, pp. 616–621.

[Mon] K. C. Mong, *Some differential invariants of 4-manifolds*, D. Phil. thesis, Oxford, 1988.

[Mai] F. Maier, *On the diffeomorphism type of elliptic surfaces with cyclic fundamental group*, thesis, Tulane Univ., New Orleans, LA, 1987.

[Ma] A. M. Macbeath, *The classification of non euclidean plane crystallographic groups*, Canad. J. Math. **19** (1967), 1199–1205.

[N] I. Nakamura, *On surfaces of Class VII_0 with curves*, Invent. Math. **78** (1984), 393–443.

[NR] W. D. Neumann and F. Raymond, *Seifert manifolds, plumbing, μ-invariants and orientation reversing maps*, Algebraic and Geometric Topology, Lecture Notes in Math., vol. 664, Springer-Verlag, Berlin and New York, 1978, pp. 162–196.

[O] C. Okonek, *Fake Enriques surfaces*, Topology **27** (1988), 415–427.

[OV1] C. Okonek and A. Van de Ven, *Stable vector bundles and differentiable structures on certain elliptic surfaces*, Invent. Math. **86** (1986), 357–370.

[OV2] _____, *Γ-type invariants associated to PU(2) bundles and the differentiable structures of Barlow's surface*, Invent. Math. **95** (1989), 601–614.

[OR] P. Orlik and F. Raymond, *Actions of the torus on 4-manifolds*. I, Trans. Amer. Math. Soc. **152** (1970), 531–559; II, Topology **13** (1974), 89–112.

[Or] P. Orlik, *Seifert manifolds*, Lecture Notes in Math., vol. 291, Springer-Verlag, Berlin and New York, 1972.

[P] P. S. Pao, *The topological structure of 4-manifolds with effective torus actions*. I, Trans. Amer. Math. Soc. **227** (1977), 279–317; II, Illinois J. Math. **21** (1977), 883–894.

[S] M. Salvetti, *On the number of non-equivalent differentiable structures on 4 manifolds*, Manuscripta Math. **63** (1989), 157–171.

[Sc] P. Scott, *The geometry of 3-manifolds*, Bull. London Math. Soc. **15** (1983), 401–487.

[Su1] T. Suwa, *On ruled surfaces of genus 1*, J. Math. Soc. Japan **21** (1969), 291–311.

[Su2] _____, *On hyperelliptic surfaces*, J. Fac. Sci. Univ. Tokyo **16** (1970), 469–476.

[SF] K. Sakamoto and S. Fukuhara, *Classification of T^2-bundles over T^2*, Tokyo J. Math. **6** (1983), 311–327.

[T] W. Thurston, *The geometry and topology of 3-manifolds*, Princeton Univ. Press (to appear).

[Tyu] A. N. Tyurin, *Algebraic geometric aspects of smooth structure*. I. *The Donaldson polynomials*, Russian Math. Surveys **44:3** (1989), 113–178.

[U1] M. Ue, *Splitting singular fibers in good torus fibrations*, J. Fac. Sci. Univ. Tokyo **32** (1985), 165–204.

[U2] _____, *On the diffeomorphism types of elliptic surfaces with multiple fibers*, Invent. Math. **84** (1986), 633–643.

[U3] _____ , *On the 4-dimensional Seifert fiberings with euclidean base orbifolds*, A Fête of Topology, Academic Press, Boston, MA, 1988, pp. 471–524.
[U4] _____ , *Geometric 4-manifolds in the sense of Thurston and Seifert 4-manifolds*. I, J. Math. Soc. Japan **42** (1990), 511–540; II, J. Math. Soc. Japan **43** (1991), 149–183.
[U5] _____ , *On the deformations of the geometric structures on the Seifert 4-manifolds*, Adv. Stud. Pure Math., vol. 20, Aspects of Low Dimensional Manifolds, 1992, pp. 331–363.
[U6] _____ , *On the decompositions of elliptic surfaces*, Knots: Proc. Internat. Conf. on Knot Theory and Related Topics (1990), de Gruyter, 1992, pp. 299–322.
[U7] _____ , *A remark on the simple invariants for elliptic surfaces and their exotic structures not coming from complex surfaces*, preprint, 1991.
[W1] C. T. C. Wall, *On simply connected 4-manifolds*, J. London Math. Soc. **39** (1964), 141–149.
[W2] _____ , *Diffeomorphisms of 4-manifolds*, J. London Math. Soc. **39** (1964), 131–140.
[W3] _____ , *Geometric structures on compact complex surfaces*, Topology **25** (1986), 119–153.
[W4] _____ , *Geometries and geometric structures in real dimension* 4 *and complex dimension* 2, Geometry and Topology, Lecture Notes in Math., vol. 1167, Springer-Verlag, Berlin and New York, 1985, pp. 268–292.
[Wo] J. A. Wolf, *Spaces of constant curvature*, 5th ed., Publish or Perish, Washington, D.C., 1984.
[Z] H. Zieschang, *On toric fiberings over surfaces*, Math. Notes **5** (1967), 341–345.
[Zimm] B. Zimmermann, *Zur Klassifikation höherdimensionaler Seifertscher Faserräume*, London Math. Soc. Lecture Note Ser., vol. 95, Cambridge Univ. Press, Cambridge, 1985, pp. 214–255.
[ZVC] H. Zieschang, E. Vogt, and H. D. Coldeway, *Surfaces and planar discontinuous groups*, Lecture Notes in Math., vol. 835, Springer-Verlag, Berlin and New York, 1980.
[Ba] S. Bauer, *Some nonreduced moduli of bundles and Donaldson invariants for Dolgachev surfaces*, J. Reine Angew. Math. **424** (1992), 149–180.
[FM4] R. Friedman and J. Morgan, *Smooth four-manifolds and complex surfaces*, Ergeb. Math. Grenzgeb., Springer-Verlag, Berlin and New York, 1994.
[EO1] W. Ebeling and C. Okonek, *Donaldson invariants, monodromy groups, and singularities*, Internat. J. Math. **1** (1990), 233–250.
[EO2] _____ , *On the diffeomorphism groups of certain algebraic surfaces*, Enseign. Math. **37** (1991), 249–262.

INSTITUTE OF MATHEMATICS, YOSHIDA COLLEGE, KYOTO UNIVERSITY, KYOTO 606, JAPAN

Translated by MASAAKI UE

Modular p-adic L-functions and p-adic Hecke Algebras

Haruzo Hida

§0

In this short article we discuss how the number of variables of modular L-functions is determined by the data from the algebraic group on which the modular forms are defined. To give you an idea about the L-functions before going into this main topic, we start with an interesting story of Euler about the values of the Riemann zeta function. This story provides a good introduction to the subject. The Riemann zeta function is defined by the following infinite series, which is absolutely convergent on the right half of the complex plane defined by $\mathrm{Re}(s) > 1$:

$$\zeta(s) = \frac{1}{1^s} + \frac{1}{2^s} + \frac{1}{3^s} + \cdots + \frac{1}{n^s} + \cdots = \sum_{n=1}^{\infty} \frac{1}{n^s}.$$

This series must be a result of a naive question: What number can one get out of the sum of negative powers of all natural numbers? Then it is natural to think about the sum of positive powers of all natural numbers. Of course, one cannot compute the outrageous sum: $1^k + 2^k + 3^k + \cdots + n^k + \cdots$ of growing integers without some trick. Euler proposed the following trick to compute this sum in the mid 18th century. Euler first supposed that the value $\zeta(-k)$ ($0 \leq k \in \mathbf{Z}$) actually exists. Then one is forced to have the following interesting identity:

$$\begin{aligned}
(1 - 2^{k+1})\zeta(-k) &= \zeta(-k) - 2^{k+1}\zeta(-k) \\
&= (1^k + 2^k + 3^k + \cdots + n^k + \cdots) \\
&\quad - 2((2 \cdot 1)^k + (2 \cdot 2)^k + (2 \cdot 3)^k + \cdots + (2 \cdot n)^k + \cdots) \\
&= \{t - 2^k t^2 + 3^k t^3 - 4^k t^4 + \cdots + (-1)^{n+1} n^k t^n + \cdots\}|_{t=1}.
\end{aligned}$$

When $k = 0$, we get

$$\begin{aligned}
-\zeta(0) &= t\{1 - t + t^2 - t^3 + \cdots + (-t)^n + \cdots\}|_{t=1} \\
&= \left.\frac{t}{1+t}\right|_{t=1} \quad \text{(the sum of a geometric series)}.
\end{aligned}$$

1991 *Mathematics Subject Classification.* Primary 11G40.
This article originally appeared in Japanese in Sūgaku **44** (4) (1992), 289–305.
The author is partially supported by an NSF grant.

This would tell us that $\zeta(0) = -\frac{1}{2}$. In general, by the well-known formula: $t\frac{d}{dt}t^n = nt^n$, we have that

(1) $$Z(-k) = (1 - 2^{k+1})\zeta(-k) = \left(t\frac{d}{dt}\right)^k \left(\frac{t}{1+t}\right)\bigg|_{t=1}.$$

Euler was not playing an empty game, and he proved the following striking formula for m a positive integer:

(2) $$\zeta(2m) = (-1)^m \frac{(2\pi)^{2m}}{2(2m-1)!}\zeta(1-2m)$$
$$= (1 - 2^{2m})^{-1}(-1)^m \frac{(2\pi)^{2m}}{2(2m-1)!} \left(t\frac{d}{dt}\right)^{2m-1}\left(\frac{t}{1+t}\right)\bigg|_{t=1}.$$

The left-hand side is the value of an actually converging infinite series and is the value of the complex analytic function $\zeta(s)$ which is well defined on the right half-plane given by $\mathrm{Re}(s) > 1$. The main part of the right-hand side is the value at $t = 1$ of the derivative of a simple rational function $t/(1+t)$. Thus, this is a formula connecting two fundamentally different objects, which was discovered by Euler almost 250 years ago. This formula looks even magical. To the author, the fun in studying number theory lies in learning and finding this sort of "magical relation" connecting two (or more) fundamentally different objects.

From today's view point, the Riemann zeta function has an analytic continuation to $\mathbf{C} - \{1\}$ having a simple pole at $s = 1$ (as proved by Riemann); the formula of Euler for $\zeta(-k)$ actually gives the value at $s = -k$ of the analytic continuation, and the formula (2) is a special case of the functional equation (proved by Riemann):

$$\zeta(s) = \frac{(2\pi)^s \zeta(1-s)}{2\Gamma(s)\cos(\pi s/2)}.$$

We refer to [W84] for historical matters about Euler and to [K75] and [H93b, §§2.1–2.2] for proofs of the above facts.

§1

There is another example of number-theoretic equality. Looking into the formula of Euler, Kummer found the following fact in the 19th century. He fixed a prime p (which we assume to be odd for simplicity). Then for two nonnegative integers k and k'

(3) $\quad k \equiv k' \bmod p^n(p-1) \Rightarrow (1 - p^k)Z(-k) \equiv (1 - p^{k'})Z(-k') \bmod p^{n+1}.$

In other words, as long as k and k' stay in the same residue class modulo $p - 1$, and if k and k' are close under the p-adic topology, then the values $(1 - p^k)Z(-k)$ and $(1 - p^{k'})Z(-k')$ are close to the same extent. This implies immediately that the function $k \mapsto (1 - p^k)Z(-k)$ for positive integers $k \equiv -1 \bmod (p-1)$ extends to a continuous function defined on the p-adic integer ring $\mathbf{Z}_p$ having values in $\mathbf{Z}_p$. We write this function as Z_p (i.e., $Z_p(-k) = (1 - p^k)Z(-k)$).

We give here a brief explanation of p-adic numbers. The reader who knows the subject well can skip this paragraph. For each integer n, we define its p-adic absolute value to be $|n|_p = p^{-e}$ if p^e divides n exactly (i.e., n/p^e is an integer but n/p^{e+1} is not an integer). We simply put $|0|_p = 0$. Then it is easy to check that this norm has all the basic properties of the usual absolute value:

(i) $|x|_p = 0 \Leftrightarrow x = 0$,
(ii) $|xy|_p = |x|_p |y|_p$,

and

(iii) $|x + y|_p \leq \max(|x|_p, |y|_p)$.

Then the function $\rho(x, y) = |x - y|_p$ is a metric and gives a structure of a metric space to $\mathbf{Z}$. For example, if $p = 5$, 5 is close to 0 ($|5|_5 = \frac{1}{5}$), $25 = 5^2$ is closer to 0 ($|25|_5 = \frac{1}{25}$) and $125 = 5^3$ is extremely close to 0 ($|125|_5 = \frac{1}{125}$). Thus, this p-adic space is an outrageous world hard to imagine for us living in the Euclidean world. We take the completion $\mathbf{Z}_p$ of $\mathbf{Z}$ under this metric (see [H93b, §1.3] for details of p-adic numbers). The ring $\mathbf{Z}_p$ is called the p-adic integer ring. We write $\mathbf{Q}_p$ for the field of fractions of $\mathbf{Z}_p$. Then $\mathbf{Q}_p$ naturally contains the rational numbers and

$$p^n \mathbf{Z}_p = \{p^n z \mid z \in \mathbf{Z}_p\} = \left\{ z \in \mathbf{Q}_p \,\middle|\, |z|_p \leq p^{-n} \right\} : \quad \text{the disk of radius } p^{-n}.$$

The ring $\mathbf{Z}_p$ is then a closed unit disk in $\mathbf{Q}_p$, and hence, is a compact ring. The fact in (3) then implies that $|Z_p(-k) - Z_p(-k')| \leq |k - k'|_p$ as long as k and k' belong to the set of integers $I = \{n \mid n \equiv -1 \bmod(p-1) \text{ and } 0 < n \in \mathbf{Z}\}$. It is easy to check that I is a dense subset of $\mathbf{Z}_p$ (under the p-adic topology). Thus, $Z_p : I \to \mathbf{Q}$ extends to a continuous function on $\mathbf{Z}_p$ having values in $\mathbf{Q}_p$. We can verify that Z_p has values in $\mathbf{Z}_p$ using formula (1). The binomial polynomial

$$\binom{s}{n} = \begin{cases} \dfrac{s(s-1)\cdots(s-n+1)}{n!} & \text{if } n > 0, \\ 1 & \text{if } n = 0 \end{cases}$$

is a polynomial with rational coefficients. This can be thought of as a polynomial function of s on $\mathbf{Z}_p$ having values in $\mathbf{Z}$ on the set of all positive integers. Then by continuity, the values $\binom{s}{n}$ fall in $\mathbf{Z}_p$ for all $s \in \mathbf{Z}_p$, i.e., $|\binom{s}{n}|_p \leq 1$. Thus, the binomial expansion

$$(1+z)^s = \sum_{n=0}^{\infty} \binom{s}{n} z^n$$

converges in $\mathbf{Z}_p$ for all $s \in \mathbf{Z}_p$ if $|z|_p < 1$. That is, on the multiplicative group

$$W = \left\{ z \in \mathbf{Z}_p \,\middle|\, |z - 1|_p < 1 \right\} = 1 + p\mathbf{Z}_p \subset \mathbf{Z}_p^{\times}$$

the p-adic power $z^s = (1 + (z-1))^s = \sum_{n=0}^{\infty} \binom{s}{n}(z-1)^n$ is a well-defined p-adic analytic function in s. Here the "p-adic analyticity" means that the function can be expanded into an absolutely convergent power series at every $s \in \mathbf{Z}_p$. This p-adic power satisfies all the formal properties of the usual complex power: $z^{s+t} = z^s z^t$, $z^0 = 1$, and so on. As is well known, the series $\{z^{p^n}\}_n$ is a p-adic Cauchy sequence and obviously $\zeta = \omega(z) = \lim_{n \to \infty} z^{p^n} \in \mathbf{Z}_p$ satisfies $\zeta^p = \zeta$. Thus, ω is a character of $\mathbf{Z}_p^{\times}$ having values in the subgroup μ (of $\mathbf{Z}_p^{\times}$) of the $(p-1)$th roots of unity. It is also easy to check $\mathrm{Ker}(\omega) = W$ (here we remind the reader that we always assume p to be odd). Thus, we can define a canonical projection $z \mapsto \langle z \rangle$ of $\mathbf{Z}_p^{\times}$ onto W by $\langle z \rangle = \omega(z)^{-1} z$. Then we can think of the p-adic analytic function $s \mapsto \langle z \rangle^s$, and

$\langle z \rangle^n = z^n$ if $n \equiv 0 \bmod(p-1)$ for integers n. Then we define the p-adic function $\zeta_p(s)$ $(s \in \mathbf{Z}_p - \{1\})$ by

(4) $$\zeta_p(-s) = (1 - \langle 2 \rangle^{s+1})^{-1} Z_p(-s).$$

Obviously this function ζ_p has a singularity at $s = 1$. From (1) we know the following striking formula due basically to Kummer:

(5) $$\zeta_p(k) = (1 - p^k)\zeta(-k) \quad \text{for all positive } k \equiv -1 \bmod(p-1).$$

Later it was shown by Kubota and Leopoldt that $\zeta_p(s)$ is actually a p-adic analytic function defined on $\mathbf{Z}_p - \{1\}$ and has a simple pole at $s = 1$. Moreover, Iwasawa showed the existence of a power series $\Phi \in \mathbf{Z}_p[[T]]$ such that $\zeta_p(s) = (u^s - 1)^{-1} \Phi(u^s - 1)$ for $u = 1 + p$ (which is a topological generator of $W \cong \mathbf{Z}_p$). A p-adic analytic function $f(s_1, \ldots, s_n)$ is called an Iwasawa function if there exists a power series Φ in $A[[X_1, \ldots, X_n]]$ for a p-adically complete valuation ring A over $\mathbf{Z}_p$ such that $f(s_1, \ldots, s_n) = \Phi(u^{s_1} - 1, \ldots, u^{s_n} - 1)$. Thus, $(u^s - 1)\zeta_p(s)$ is an Iwasawa function. The formula (5) connects the complex analytic function $\zeta(s)$ and the p-adic analytic function $\zeta_p(s)$ at positive integers $k \equiv -1 \bmod(p-1)$. Then it is natural to call ζ_p the p-adic Riemann zeta function. We again refer to [**H93b**, §§3.4–3.5] for the proof of these facts.

§2

Now we enter into the principal subject of the note. Although the L-functions we often encounter in number theory are functions of one complex or p-adic variable, there is no apparent reason not to have L-functions of several variables. Probably, we number theorists are too familiar with one-variable L-functions to notice the following fundamental question:

Why do L-functions have only one variable?

An explanation to this fact seems to have been offered first by A. Weil. To explain this, we consider the adèle ring $\mathbf{A}$ of $\mathbf{Q}$. The formal definition of $\mathbf{A}$ is as follows: $\mathbf{A}$ is the subring of $\mathbf{R} \times \prod_p \mathbf{Q}_p$ generated by the diagonal image of $\mathbf{Q}$ and $\widehat{\mathbf{Z}} \times \mathbf{R}$, where $\widehat{\mathbf{Z}} = \prod_p \mathbf{Z}_p$. Here p runs over all primes including 2. The field $\mathbf{Q}$ is considered to be a subring of the product $\mathbf{R} \times \prod_p \mathbf{Q}_p$ via the diagonal map $\mathbf{Q} \ni \alpha \mapsto (\alpha, \ldots, \alpha, \alpha, \ldots) \in \mathbf{R} \times \prod_p \mathbf{Q}_p$. The ring $\widehat{\mathbf{Z}}$ is a product of compact rings $\mathbf{Z}_p$, and hence, is compact. As is easily seen, $\mathbf{A} = \mathbf{Q} + \widehat{\mathbf{Z}} \times \mathbf{R}$ in the product $\mathbf{R} \times \prod_p \mathbf{Q}_p$. Since $\widehat{\mathbf{Z}} \times \mathbf{R} \cap \mathbf{Q} = \mathbf{Z}$, we know that

$$\mathbf{A}/\widehat{\mathbf{Z}} \times \mathbf{R} \cong \mathbf{Q}/\mathbf{Z}.$$

We put on $\widehat{\mathbf{Z}} \times \mathbf{R}$ the product topology of the compact ring $\widehat{\mathbf{Z}}$ and the number line $\mathbf{R}$. Then we extend this topology on $\widehat{\mathbf{Z}} \times \mathbf{R}$ to $\mathbf{A}$ by translation by elements in $\mathbf{Q}$. Thus $\mathbf{A}$ is a locally compact ring. Elements in $\mathbf{A}$ are called *adèles* (see [**H93b**, §8.1] for details of adèles). In particular, we look into the multiplicative group $\mathbf{A}^\times$ of this ring whose elements are called *idèles*. If $z = (z_p)$ is an idèle, then z can be written as $\alpha + z_0$ with $z_0 \in \widehat{\mathbf{Z}} \times \mathbf{R}$ and $\alpha \in \mathbf{Q}$. Thus, there are only finitely many primes p such that $|z_p|_p \neq 1$. By definition, $|z_p|_p$ is a power of p such that $\big||z_p|_p\big|_p = 1$. Thus, the rational number $\alpha = \prod_p |z_p|_p$ is well defined because $|z_p|_p = 1$ for all but finitely many primes p. Then $|\alpha z_p|_p = 1$, and thus, $\alpha z_p \in \mathbf{Z}_p^\times$. We can also replace α by $-\alpha$, if necessary, to ensure that $\alpha z_\infty \in \mathbf{R}_+^\times$, where z_∞ denotes the component of z in $\mathbf{R}$ and $\mathbf{R}_+^\times$ is the group of all positive real numbers. This implies $\alpha z \in \widehat{\mathbf{Z}}^\times \times \mathbf{R}_+^\times$, and thus, we have a natural surjective group homomorphism $\iota \colon \widehat{\mathbf{Z}}^\times \to \mathbf{A}^\times / \mathbf{Q}^\times \mathbf{R}_+^\times$, $(z \mapsto (z, 1))$. It is then immediate that

$$\widehat{\mathbf{Z}}^\times \cap \mathbf{Q}^\times \mathbf{R}_+^\times = \{1\}.$$

Thus, ι is an isomorphism $\widehat{\mathbf{Z}}^\times \cong \mathbf{A}^\times / \mathbf{Q}^\times \mathbf{R}_+^\times$. Defining the adèle norm $|z|_\mathbf{A}$ by $|z_\infty| \times \prod_p |z|_p$, we also have from the above argument the product formula

$$|\alpha|_\mathbf{A} = 1 \quad \text{for all } \alpha \in \mathbf{Q}.$$

The notion of adèles was created basically to supply a good tool to describe class field theory. We fix an algebraic closure $\overline{\mathbf{Q}}$ of $\mathbf{Q}$. We may identify $\overline{\mathbf{Q}}$ with the totality of all roots in $\mathbf{C}$ of polynomial equations with coefficients in $\mathbf{Q}$. A Galois extension K/F in $\overline{\mathbf{Q}}$ is called abelian if $\mathrm{Gal}(K/F)$ is an abelian group. Abelian extensions have a remarkable property that the composite of any two abelian extensions is again abelian. Thus, there exists the maximal abelian extension written $F_{\mathrm{ab}} (\subset \overline{\mathbf{Q}})$ over F which is the composite of all abelian extensions of F. The extension F_{ab} is the fixed field of the commutator subgroup of $\mathrm{Gal}(\overline{\mathbf{Q}}/F)$. A typical example of an abelian extension of $\mathbf{Q}$ is the field generated by all Nth roots of unity for any given integer $N > 1$. If we write $\zeta = \exp(2\pi\sqrt{-1}/N)$, then the field is just $\mathbf{Q}(\zeta)$. Note that

$$\mu_N = \{x \in \overline{\mathbf{Q}}^\times \mid x^N = 1\} = \{\zeta^n \mid n \in (\mathbf{Z}/N\mathbf{Z})\}$$
$$\cong (\mathbf{Z}/N\mathbf{Z}): \quad \text{cyclic group of order } N.$$

For each automorphism σ of $\mathbf{Q}(\zeta)$, σ^{-1} takes ζ to another Nth root of unity $\zeta^{\alpha(\sigma)}$ $(\alpha(\sigma) \in \mathbf{Z}/N\mathbf{Z})$. Since $\zeta^{\alpha(\sigma)}$ is again a generator of the cyclic group μ_N, $\alpha(\sigma)$ belongs to the unit group $(\mathbf{Z}/N\mathbf{Z})^\times$. Naturally, $\alpha(\sigma\tau) = \alpha(\sigma)\alpha(\tau)$, and σ is determined by the value $\alpha(\sigma)$. It is well known that $\alpha \colon \mathrm{Gal}(\mathbf{Q}(\zeta)/\mathbf{Q}) \to (\mathbf{Z}/N\mathbf{Z})^\times$ is a surjective isomorphism. Thus, $\mathbf{Q}_{\mathrm{ab}}$ contains Nth roots of unity for all n. We can identify the group μ_∞ of all roots of unity with $\mathbf{Q}/\mathbf{Z} = \mathbf{A}/\widehat{\mathbf{Z}} \times \mathbf{R}$ via $\mathbf{Q} \ni a \mapsto \exp(2\pi i a) \in \mu_\infty$. Although $\mathbf{Q}_{\mathrm{ab}}$ is an infinite extension of $\mathbf{Q}$, we can still think of the huge abelian group $\mathrm{Gal}(\mathbf{Q}_{\mathrm{ab}}/\mathbf{Q})$. Observing that $\mathrm{Aut}(\mathbf{A}/\widehat{\mathbf{Z}} \times \mathbf{R}) = \widehat{\mathbf{Z}}^\times$ via multiplication by elements of $\widehat{\mathbf{Z}}^\times$, we have a group homomorphism, called the cyclotomic character:

(6) $\quad \alpha \colon \mathrm{Gal}(\mathbf{Q}_{\mathrm{ab}}/\mathbf{Q}) \to \mathrm{Aut}(\mu_\infty) = \mathrm{Aut}(\mathbf{A}/\widehat{\mathbf{Z}} \times \mathbf{R}) = \widehat{\mathbf{Z}}^\times \cong \mathbf{A}^\times / \mathbf{Q}^\times \mathbf{R}_+^\times$

given by $\zeta^\sigma = \zeta^{\alpha(\sigma)}$. Class field theory claims that α is in fact a surjective isomorphism:

$$\mathrm{Gal}(\mathbf{Q}_{\mathrm{ab}}/\mathbf{Q}) \cong \mathbf{A}^\times / \mathbf{Q}^\times \mathbf{R}_+^\times.$$

This fact was actually conceived, before the appearance of class field theory, by Kronecker as his famous theorem asserting that $\mathbf{Q}_{\mathrm{ab}}$ is generated by roots of unity. The analogous fact is true for any number field (i.e., finite extensions of $\mathbf{Q}$). That is, defining a locally compact ring $F_\mathbf{A} = F \otimes_\mathbf{Q} \mathbf{A}$ ($\cong \mathbf{A}^{[F:\mathbf{Q}]}$ as topological spaces) and regarding F as a subfield of $F_\mathbf{A}$ by $F \ni a \mapsto a \otimes 1$, we have a canonical Artin reciprocity homomorphism: $F_\mathbf{A}^\times / F^\times \to \mathrm{Gal}(F_{\mathrm{ab}}/F)$ whose kernel is the connected component of the idèle class group $C_F = F_\mathbf{A}^\times / F^\times$ and which is surjective. The above identity is another number-theoretic identity connecting quite different objects.

For any infinite Galois extension $K/\mathbf{Q}$, there is a natural topology on $\mathrm{Gal}(K/\mathbf{Q})$, called the Krull topology. Its fundamental system of neighborhoods of 1 is given by the set of all subgroups that fix a finite extension of $\mathbf{Q}$. Under this topology, $\mathrm{Gal}(K/\mathbf{Q})$ is a compact group (see [N86, 1.1]). Then the isomorphism α is in fact an isomorphism of two compact groups.

Since $\mathrm{Gal}(\mathbf{Q}_{\mathrm{ab}}/\mathbf{Q})$ is totally disconnected, it does not have many nontrivial complex (quasi) characters. However, the usefulness of characters for studying a given group is something everyone knows by experience. In particular, complex characters are easy to handle. Class field theory offers a way to get meaningful complex characters. The idea is to inflate the Galois group giving a nontrivial connected component. We consider $\mathbf{A}^\times/\mathbf{Q}^\times$ instead of $\mathrm{Gal}(\mathbf{Q}_{\mathrm{ab}}/\mathbf{Q}) = \mathbf{A}^\times/\mathbf{Q}^\times \mathbf{R}_+^\times$. Then we have a nontrivial group of quasicharacters:
$$\mathrm{Hom}_{\mathrm{conti}}(\mathbf{A}^\times/\mathbf{Q}^\times, \mathbf{C}^\times).$$
There are infinitely many connected components in the space $\mathrm{Hom}_{\mathrm{conti}}(\mathbf{A}^\times/\mathbf{Q}^\times, \mathbf{C}^\times)$ which is represented by finite-order characters ξ of $\mathbf{A}^\times/\mathbf{Q}^\times$. The connected component containing ω is isomorphic to $\mathbf{C}$ via $\mathbf{C} \ni s \mapsto \xi(x)|x|_\mathbf{A}^s$. In particular, we look at the principal component corresponding to the trivial ξ:
$$\mathrm{Hom}_{\mathrm{conti}}(\mathbf{A}^\times/\widehat{\mathbf{Z}}^\times \mathbf{Q}^\times, \mathbf{C}^\times) = \{|x|_\mathbf{A}^s \mid s \in \mathbf{C}\} \cong \mathbf{C}.$$

For any idèle z, we define its finite part z_f by zz_∞^{-1}. Thus, z_f has the same component as z at each prime p, but its infinite component is equal to 1. Note that for each positive integer n, $|n_f|_\mathbf{A}^s = n^{-s}$, and
$$\zeta(s) = \sum_{n=1}^\infty \frac{1}{n^s} = \sum_{n=1}^\infty |n_f|_\mathbf{A}^s$$
can be considered to be a function on the principal connected component
$$\mathrm{Hom}_{\mathrm{conti}}(\mathbf{A}^\times/\widehat{\mathbf{Z}}^\times \mathbf{Q}^\times, \mathbf{C}^\times)$$
of $\mathrm{Hom}_{\mathrm{conti}}(\mathbf{A}^\times/\mathbf{Q}^\times, \mathbf{C}^\times)$ corresponding to the identity character. This is the reason that $\zeta(s)$ has only one complex variable. Note that $\mathbf{A}^\times/\widehat{\mathbf{Z}}^\times \mathbf{Q}^\times \cong \mathbf{R}_+^\times$ because $\mathbf{A}^\times = \widehat{\mathbf{Z}}^\times \mathbf{Q}^\times \mathbf{R}_+^\times$ and $\widehat{\mathbf{Z}}^\times \mathbf{Q}^\times \cap \mathbf{R}_+^\times = \{1\}$. Thus, we again obtain
$$\mathrm{Hom}_{\mathrm{conti}}(\mathbf{A}^\times/\widehat{\mathbf{Z}}^\times \mathbf{Q}^\times, \mathbf{C}^\times) \cong \mathrm{Hom}_{\mathrm{conti}}(\mathbf{R}_+^\times, \mathbf{C}^\times) \cong \mathbf{C} \quad \text{via } (x \mapsto x^s) \leftrightarrow s$$
which is the domain on which $\zeta(s)$ is defined.

§3

We now argue in the same way to show that the p-adic Riemann zeta function necessarily has only one variable. When we considered the complex zeta function $\zeta(s)$, we removed the factor $\mathbf{R}_+^\times$ at infinity from the denominator of $\mathbf{A}^\times/\mathbf{Q}^\times \mathbf{R}_+^\times = \mathrm{Gal}(\mathbf{Q}_{\mathrm{ab}}/\mathbf{Q})$ to obtain the right group on which L-functions are defined. Since the domain on which $\zeta(s)$ is defined is given by $\mathrm{Hom}_{\mathrm{conti}}(\mathbf{A}^\times/\widehat{\mathbf{Z}}^\times \mathbf{Q}^\times, \mathbf{C}^\times)$, we remove the factor $\mathbf{Z}_p^\times$ at p from the denominator $\widehat{\mathbf{Z}}^\times \mathbf{Q}^\times$ and insert $\mathbf{R}_+^\times$ to fill the infinity component. Then we expect to have $\mathrm{Hom}_{\mathrm{conti}}(\mathbf{A}^\times/U^{(p)}\mathbf{Q}^\times, \mathbf{Q}_p^\times)$ as the domain of the p-adic Riemann zeta function, where $U^{(p)} = \mathbf{R}_+^\times \times \prod_{l \neq p} \mathbf{Z}_l^\times$. Note here that the group $\mathbf{A}^\times/U^{(p)}\mathbf{Q}^\times$ is a quotient group of $\mathbf{A}^\times/\widehat{\mathbf{Z}}^\times \mathbf{Q}^\times = \mathrm{Gal}(\mathbf{Q}_{\mathrm{ab}}/\mathbf{Q})$. Therefore, there is a subfield X of $\mathbf{Q}_{\mathrm{ab}}$ such that $\mathrm{Gal}(X/\mathbf{Q}) \cong \mathbf{A}^\times/U^{(p)}\mathbf{Q}^\times$. The field X is given by $\mathbf{Q}(1^{1/p^\infty})$ which is the field generated by all p-power roots of unity as we see from the construction of α in (6). Thus, we have the cyclotomic character

$$(7) \qquad \chi: \mathbf{A}^\times/U^{(p)}\mathbf{Q}^\times \cong \mathrm{Gal}(\mathbf{Q}(1^{1/p^\infty})/\mathbf{Q}) \cong \mathbf{Z}_p^\times,$$

which satisfies $\chi(z) = z$ for $z \in \mathbf{Z}_p^\times$. Since $\mathbf{Z}_p^\times$ is a compact group (it is a closed disk of radius p^{-1} centered at 1), the image of $\mathbf{Z}_p^\times$ under the continuous homomorphism lies in the maximal compact subgroup of $\mathbf{Q}_p^\times$, which is again $\mathbf{Z}_p^\times$. For each finite group G, writing $A[G]$ for the group algebra of G over a given commutative ring A, we know that

$$\mathrm{Hom}_{\mathrm{gr}}(G, A^\times) \cong \mathrm{Hom}_{A\text{-alg}}(A[G], A).$$

Similarly, any continuous homomorphism $\phi: \mathbf{Z}_p^\times \to \mathbf{Z}_p^\times$ induces modulo p^n a group homomorphism $\phi_n: \mathbf{Z}_p^\times \to (\mathbf{Z}/p^n\mathbf{Z})^\times$ such that $\phi_m \bmod p^n = \phi_n$ for all $m > n$. Note here that $\mathbf{Z}_p^\times = W \times \mu$, $\mathbf{Z}_p \cong W$ via $s \mapsto u^s$ ($u = 1 + p$), and $(\mathbf{Z}/p^n\mathbf{Z})^\times$ is a cyclic group of order $p^{n-1}(p-1)$ (here we still assume that p is odd). Thus, any $x \in W^{p^{n-1}}$ is $y^{p^{n-1}(p-1)}$ for some $y \in \mathbf{Z}_p^\times$. That is, $\phi_n(x) = 1$ for all $x \in W^{p^{n-1}}$. Hence, ϕ_n must factor through $\mathbf{Z}_p^\times/W^{p^{n-1}} \cong (\mathbf{Z}/p^n\mathbf{Z})^\times$. As a consequence of this argument, we have

$$\mathrm{Hom}_{\mathrm{conti}}(\mathbf{Z}_p^\times, \mathbf{Z}_p^\times) = \varprojlim_n \mathrm{Hom}_{\mathrm{gr}}((\mathbf{Z}/p^n\mathbf{Z})^\times, (\mathbf{Z}/p^n\mathbf{Z})^\times)$$

$$= \varprojlim_n \mathrm{Hom}_{\mathbf{Z}_p\text{-alg}}(\mathbf{Z}_p[(\mathbf{Z}/p^n\mathbf{Z})^\times], (\mathbf{Z}/p^n\mathbf{Z}))$$

$$= \mathrm{Hom}_{\mathbf{Z}_p\text{-alg}}(\mathbf{Z}_p[[\mathbf{Z}_p^\times]], \mathbf{Z}_p),$$

where $\mathbf{Z}_p[[\mathbf{Z}_p^\times]] = \varprojlim_n \mathbf{Z}_p[(\mathbf{Z}/p^n\mathbf{Z})^\times]$ is called the continuous group algebra of $\mathbf{Z}_p^\times$. In general, for each profinite group $G = \varprojlim_n G_n$ for finite groups G_n, the continuous group algebra is defined by

$$\mathbf{Z}_p[[G]] = \varprojlim_n \mathbf{Z}_p[G_n],$$

where the transition map $\rho: \mathbf{Z}_p[G_m] \to \mathbf{Z}_p[G_n]$ ($m \geq n$) is obtained by $\rho(\Sigma_g a_g g) = \Sigma_g a_g \rho(g)$ from the transition map $\rho: G_m \to G_n$.

Writing μ for the group of $(p-1)$th roots of unity in $\mathbf{Z}_p^\times$, we have, as already seen, $\mathbf{Z}_p^\times = W \times \mu$. If $p - 1$ is invertible in an integral domain A (for example, in $\mathbf{Z}_p$, $(p-1)^{-1} = -(1 + p + p^2 + p^3 + \cdots) \in \mathbf{Z}_p$) and if A contains all $(p-1)$th roots of unity, then the group algebra $A[\mu]$ is isomorphic to the product of copies of A indexed by characters of μ having values in $A^\times$. We write $\hat{\mu}$ for the set of all characters of μ. Then $\hat{\mu} = \{\omega^a | a = 0, \ldots, p-2\}$ for the inclusion $\omega: \mu \to A^\times$. Each projection of $A[\mu]$ onto the component A indexed by ω^a is actually given by the algebra homomorphism corresponding to ω^a. Note that $\mathbf{Z}_p^\times/W^{p^n} = \mu \times (W/W^{p^n})$, and hence,

$$\mathbf{Z}_p[\mu \times (W/W^{p^n})] = \mathbf{Z}_p[W/W^{p^n}][\mu] = \prod_{\xi \in \hat{\mu}} \mathbf{Z}_p[W/W^{p^n}]$$

and

$$\mathbf{Z}_p[[\mathbf{Z}_p^\times]] = \prod_{\xi \in \hat{\mu}} \mathbf{Z}_p[[W]].$$

Thus, we see that

$$\mathrm{Hom}_{\mathrm{conti}}(\mathbf{Z}_p^\times, \mathbf{Z}_p^\times) = \mathrm{Hom}_{\mathbf{Z}_p\text{-alg}}(\mathbf{Z}_p[[\mathbf{Z}_p^\times]], \mathbf{Z}_p)$$
$$= \coprod_{\xi \in \hat{\mu}} \mathrm{Hom}_{\mathbf{Z}_p\text{-alg}}(\mathbf{Z}_p[[W]], \mathbf{Z}_p).$$

The last identification is given just because any $\mathbf{Z}_p$-algebra homomorphism of $\mathbf{Z}_p[[\mathbf{Z}_p^\times]]$ into $\mathbf{Z}_p$ must factor through a component $\mathbf{Z}_p[[W]]$ for a unique $\xi \in \hat{\mu}$. We know that W/W^{p^n} is a cyclic group of order p^n. Thus,

$$\mathbf{Z}_p[W/W^{p^n}] \cong \mathbf{Z}_p[t]/(t^{p^n} - 1) \cong \mathbf{Z}_p[T]/((1 + T)^{p^n} - 1) \text{ (by } t = 1 + T).$$

Then it is easy to see (cf. [Wa82, Theorem 7.1]) that

$$\Lambda = \mathbf{Z}_p[[W]] \cong \mathbf{Z}_p[[T]] = \varprojlim_n \mathbf{Z}_p[T]/((1 + T)^{p^n} - 1).$$

Thus, any continuous algebra homomorphism ϕ of Λ into $\mathbf{Z}_p$ is determined by its value at T. Since $\lim_{n\to\infty} T^{p^n} = 0$ in Λ, $\lim_{n\to\infty} \phi(t)^{p^n} = 1$, and hence, $\phi(t) = u^s \in W$. Thus,

$$\mathrm{Hom}_{\mathbf{Z}_p\text{-alg}}(\mathbf{Z}_p[[W]], \mathbf{Z}_p) \cong \mathbf{Z}_p \quad \text{via } \phi \leftrightarrow s \text{ for } \phi(T) = u^s.$$

In particular, the domain on which $\zeta_p(s)$ is defined is the component

$$\mathrm{Hom}_{\mathbf{Z}_p\text{-alg}}(\mathbf{Z}_p[[W]], \mathbf{Z}_p)$$

corresponding to the identity character $\xi = \omega^0$.

We now fix an algebraic closure $\overline{\mathbf{Q}}_p$ of $\mathbf{Q}_p$. Thus $\overline{\mathbf{Q}}$ is a subfield of $\mathbf{C}$ as well as $\overline{\mathbf{Q}}_p$. There is a natural p-adic norm $|\ |_p$ on $\overline{\mathbf{Q}}_p$ extending $|\ |_p$ on $\mathbf{Q}_p$. We can take $\mathrm{Hom}_{\mathbf{Z}_p\text{-alg,conti}}(\mathbf{Z}_p[[W]], \overline{\mathbf{Q}}_p)$ in place of $\mathrm{Hom}_{\mathbf{Z}_p\text{-alg}}(\mathbf{Z}_p[[W]], \mathbf{Z}_p)$. It is easy to see that

(8) $\quad \mathrm{Hom}_{\mathbf{Z}_p\text{-alg,conti}}(\mathbf{Z}_p[[W]], \overline{\mathbf{Q}}_p) \cong \left\{ x \in \overline{\mathbf{Q}}_p \,\Big|\, |x|_p < 1 \right\} = D \quad \text{via } \phi \leftrightarrow \phi(T).$

As we have already noted, $(u^s - 1)\zeta_p(s)$ is an Iwasawa function, i.e., $(u^s - 1)\zeta_p(s) = \Phi(u^s - 1)$ for $\Phi \in \mathbf{Z}_p[[T]]$. Thus in fact, we may regard $\zeta_p(s)$ as a function on D given by $x\zeta_p(x) = \Phi(x)$ for $x \in D$. This is the point of view taken in [I72]. In particular, the p-adic Riemann zeta function is legitimately of one variable.

§4

The principle behind our argument in §2 and §3 is to regard each L-function as a function of characters. Then the values of characters can be considered as eigenvalues of the regular representation on a suitable space of functions on $\mathbf{Q}^\times \backslash \mathbf{A}^\times = \mathrm{GL}_1(\mathbf{Q}) \backslash \mathrm{GL}_1(\mathbf{A})$. That is, for each function $\phi: \mathrm{GL}_1(\mathbf{Q}) \backslash \mathrm{GL}_1(\mathbf{A}) \to \mathbf{C}$, we define the representation R of $\mathrm{GL}_1(\mathbf{A})$ by $(R(g)\phi)(x) = \phi(xg)$. A typical choice of such spaces is the L_2-space on $\mathrm{GL}_1(\mathbf{Q}) \backslash \mathrm{GL}_1(\mathbf{A})$ with respect to the Haar measure on $\mathrm{GL}_1(\mathbf{Q}) \backslash \mathrm{GL}_1(\mathbf{A})$. Actually, for each Hecke character ξ, we may regard ξ as the eigenvector of the operators $R(g)$: $R(g)\xi = \xi(g)\xi$ with eigenvalue $\xi(g)$.

Now we explore the same question for nonabelian L-functions. The idea of Langlands is to use the topological space $\mathrm{GL}_n(\mathbf{Q}) \backslash \mathrm{GL}_n(\mathbf{A})$. The group $\mathrm{GL}_n(\mathbf{A})/\mathbf{R}_+^\times$ is a locally compact group having a Haar measure $d\mu_0(x)$. Since $\mathrm{GL}_n(\mathbf{Q})$ acts discretely on $\mathrm{GL}_n(\mathbf{A})/\mathbf{R}_+^\times$, we have a fundamental domain Φ in $\mathrm{GL}_n(\mathbf{A})/\mathbf{R}_+^\times$ of

$\mathrm{GL}_n(\mathbf{Q})$. We then define the invariant measure $d\mu$ on $X = \mathrm{GL}_n(\mathbf{Q}) \backslash \mathrm{GL}_n(\mathbf{A})/\mathbf{R}_+^\times$ by $\int_X \phi \, d\mu(x) = \int_\Phi \phi \, d\mu_0(x)$. We fix a Hecke character ξ of $\mathbf{A}^\times/\mathbf{Q}^\times$ and consider measurable functions $\phi \colon \mathrm{GL}_n(\mathbf{A}) \to \mathbf{C}$ satisfying

(I) $\phi(\alpha z x) = \xi(z)\phi(x)$ for all scalar matrices $z \in \mathbf{A}^\times$ and $\alpha \in \mathrm{GL}_n(\mathbf{Q})$.

For two functions ϕ and ϕ' satisfying (I), the function $x \mapsto \overline{\phi(x)}\phi'(x)|\xi(\det(x))|^{-1}$ is invariant under left multiplication by $\mathbf{R}_+^\times$ and right multiplication by $\mathrm{GL}_n(\mathbf{Q})$. Thus, we can define the inner product of such functions by

$$(\phi, \phi') = \int_X \overline{\phi(x)}\phi'(x)|\xi(\det(x))|^{-1} \, d\mu(x).$$

Then the L_2-space $L_2(\xi)$ is defined with respect to this inner product. We can let $\mathrm{GL}_n(\mathbf{A})$ act on $L_2(\xi)$ by $(R_\xi(g)\phi)(x) = \phi(xg)$. We thus have a representation R_ξ that satisfies $(R_\xi(g)\phi, R_\xi(g)\phi') = |\xi(\det(g))|(\phi, \phi')$. Thus, the twisted representation $R_\xi \otimes |\xi|^{-1/2} ((R_\xi \otimes |\xi|^{-1/2})(g) = |\xi(\det(g))|^{-1/2} R_\xi(g))$ is unitary and has a spectral decomposition. It is known by Langlands' theory of Eisenstein series [**L76**] that we have

$$R_\xi = \widehat{\bigoplus}_{\pi : \text{ irreducible}} \pi \oplus \left\{ \bigoplus \text{continuous sum} \right\}.$$

The continuous sum is at most $(n-1)$-dimensional. We may also consider the regular representation R on $L_2(\mathrm{GL}_n(\mathbf{Q}) \backslash \mathrm{GL}_n(\mathbf{A}))$ defined in a way similar to R_ξ. Thus, the regular representation R on $L_2(\mathrm{GL}_n(\mathbf{Q}) \backslash \mathrm{GL}_n(\mathbf{A}))$ has at most an n-dimensional continuous spectrum, and we write this fact *symbolically* as

$$R = \widehat{\bigoplus} \int_{\widehat{G}} \pi \otimes \eta \, d\eta \oplus \left\{ \bigoplus \int_{\widehat{G}} (\text{continuous sum}) \, d\eta \right\} \quad \text{for a fixed } \xi,$$

where $\widehat{G}$ is the space of unitary characters of $G = \mathbf{A}^\times/\mathbf{Q}^\times$ (which is isomorphic to the disjoint union of $\sqrt{-1}\mathbf{R}$ because $\mathrm{Hom}_{\text{conti}}(\mathbf{A}^\times/\mathbf{Q}^\times, \mathbf{C}^\times)$ is a disjoint union of $\mathbf{C}$). Now we assume that π is cuspidal, i.e., π is realized on the subspace

$$L_2^0(\xi) = \left\{ \phi \in L_2(\xi) \mid \int_{N(\mathbf{Q})\backslash N(\mathbf{A})} \phi(xn) \, dn = 0 \text{ for almost all } x \in \mathrm{GL}_n(\mathbf{A}) \right\},$$

where in the above definition N runs over all standard unipotent subgroups in $\mathrm{GL}_n(\mathbf{Q})$. The representation on the space $L_2^0(\xi)$ is known to be decomposed into a discrete sum of irreducible representations in which each irreducible representation occurs at most once. We then know that π as above can be decomposed as $\bigotimes_p \pi_p \otimes \pi_\infty$ for local representations π_p of $\mathrm{GL}_n(\mathbf{Q}_p)$ and π_∞ of $\mathrm{GL}_n(\mathbf{R})$. Moreover, for all but finitely many p, the representation π_p has a vector in its representation space fixed by $K_p = \mathrm{GL}_n(\mathbf{Z}_p)$. Such representations having a nontrivial fixed vector by $K_p = \mathrm{GL}_n(\mathbf{Z}_p)$ are called spherical (or unramified). If π_p is spherical and irreducible, then the vectors fixed by K_p form a one-dimensional space $V(\pi_p)$. We then define an action of the double coset $K_p g K_p$ for $g \in M_n(\mathbf{Z}_p) \cap \mathrm{GL}_n(\mathbf{Q}_p)$ on $V(\pi_p)$ by

$$\pi(K_p g K_p) v = \sum_{g_i} \pi(g_i) v \quad \text{for any disjoint decomposition } K_p g K_p = \coprod_i g_i K_p.$$

Since $\pi(K_p g K_p) v$ is an average of $\pi(g) v$ over a double coset $K_p g K_p$, it again falls in $V(\pi_p)$. Thus, v is an eigenvector of all operators of the form of $T = \sum_g \pi(K_p g K_p)$.

We write $\lambda(T)$ for the eigenvalue of T. We specify these operators as follows: first, decomposing the set $\{g \in M_n(\mathbf{Z}_p) \mid \det(g)\mathbf{Z}_p = p^m\mathbf{Z}_p^\times\}$ into a disjoint union of double cosets $K_p g K_p$, we define the Hecke operator T_{p^m} by $\Sigma_g \pi(K_p g K_p)$. We write $T_i(p) = \pi(K_p \varpi_i K_p)$ for the diagonal matrix ϖ_i having i p's and $n-i$ 1's as diagonal entries. Then we can show [**Sh71**, Theorem 3.21] that the formal power series $\sum_{m=0}^{\infty} \lambda(T_{p^m}) X^m$ is a rational function, and in fact, we have

$$\text{(L)} \quad \sum_{m=0}^{\infty} \lambda(T_{p^m}) X^m = L_{\pi_p}(X)^{-1} \quad \text{for } L_{\pi_p}(X) = \sum_{i=0}^{n} (-1)^i p^{i(i-1)/2} \lambda(T_i(p)) X^i.$$

We define $L(s, \pi_p \otimes \eta_p) = L_{\pi_p \otimes \eta_p}(p^{-s})^{-1}$ if $\pi \otimes \eta_p$ is spherical for each quasicharacter η_p of $\mathbf{Q}_p^\times$. There is a representation theoretic way of defining a polynomial $L_{\pi_p \otimes \eta_p}(X)$ of degree $\leq n$ with $L_{\pi \otimes \eta_p}(0) = 1$ even for nonspherical $\pi_p \otimes \eta_p$ [**GJ72**]. We define $L(s, \pi_p) = L_{\pi \otimes \eta_p}(p^{-s})^{-1}$ for such π_p and η_p, and we set

$$L(s, \pi \otimes \eta) = \prod_p L(s, \pi_p \otimes \eta_p) \quad \text{for each Hecke character } \eta \colon \mathbf{Q}^\times \backslash \mathbf{A}^\times \to \mathbf{C}^\times,$$

which converges absolutely if $\operatorname{Re}(s)$ is sufficiently large, has an analytic continuation to the entire complex plane, and satisfies a good functional equation [**GJ72**]. Thus, we may regard $L(s, \pi \otimes \eta)$ as a function defined on the connected component of the spectrum, which is isomorphic to the connected component of the space of Hecke characters $\operatorname{Hom}_{\text{conti}}(\mathbf{Q}^\times \backslash \mathbf{A}^\times, \mathbf{C}^\times)$ via $\pi \otimes \eta \leftrightarrow \eta$. Thus, $L(s, \pi \otimes \eta)$ is of one variable.

§5

When we consider the commutative algebraic group $G = \operatorname{GL}(1)$, there is a surjective homomorphism $\operatorname{GL}_1(\mathbf{Q}) \backslash \operatorname{GL}_1(\mathbf{A})$ onto $\operatorname{Gal}(\mathbf{Q}_{\text{ab}}/\mathbf{Q})$. Thus, the automorphic side $\operatorname{GL}_1(\mathbf{Q}) \backslash \operatorname{GL}_1(\mathbf{A})$ and the Galois side $\operatorname{Gal}(\mathbf{Q}_{\text{ab}}/\mathbf{Q})$ are directly linked. In the nonabelian case, the automorphic side is just a topological space $\operatorname{GL}_n(\mathbf{Q}) \backslash \operatorname{GL}_n(\mathbf{A})$, and at the Galois side there is no way to define naively an extension $\mathbf{Q}_{\operatorname{GL}(n)}$ of $\mathbf{Q}$ which replaces $\mathbf{Q}_{\text{ab}} = \mathbf{Q}_{\operatorname{GL}(1)}$ in the sense that the irreducible factors of the regular representation of $\operatorname{GL}_n(\mathbf{A})$ on $L_2(\operatorname{GL}_n(\mathbf{Q}) \backslash \operatorname{GL}_n(\mathbf{A}))$ correspond to the irreducible representations of $\operatorname{Gal}(\mathbf{Q}_{\operatorname{GL}(n)}/\mathbf{Q})$. Thus, the automorphic side and the Galois side seem to come apart in the nonabelian case.

Even in the abelian case, we need to replace the actual Galois group by $\mathbf{Q}^\times \backslash \mathbf{A}^\times$ to get many complex quasicharacters. Thus, we need to find a good group extending $\operatorname{Gal}(\overline{\mathbf{Q}}/\mathbf{Q})$ for the Galois side. The first attempt to find such a group was made by A. Weil. He found, up to isomorphism, a system of groups called the Weil groups [**W51**] (see also [**T79**]): To get $\mathbf{Q}^\times \backslash \mathbf{A}^\times$, we extended the Galois group $\operatorname{Gal}(\mathbf{Q}_{\text{ab}}/\mathbf{Q})$ by the connected component of $\mathbf{Q}^\times \backslash \mathbf{A}^\times$. Thus, to get the nonabelian version of $\mathbf{Q}^\times \backslash \mathbf{A}^\times$, we need to inflate the total Galois group $\operatorname{Gal}(\overline{\mathbf{Q}}/\mathbf{Q})$ by a suitable connected component which has to be compatible with class field theory over all finite extensions $F/\mathbf{Q}$. Thus, the Weil groups W_F are indexed by finite extensions $F/\mathbf{Q}$ and are in fact a system of triples $(W_F, \varphi, \{r_F\})$ consisting of a locally compact topological group W_F satisfying $W_F \supset W_E$ if $E \supset F$, a surjective homomorphism $\varphi \colon W_\mathbf{Q} \to \operatorname{Gal}(\overline{\mathbf{Q}}/\mathbf{Q})$ of topological groups inducing $W_F/W_E \cong \operatorname{Hom}_F(E, \overline{\mathbf{Q}})$ *for all finite extensions* $E/F/\mathbf{Q}$ and another isomorphism of topological groups $r_E \colon C_E \cong W_E^{\text{ab}} = W_E/\overline{[W_E \colon W_E]}$ for each finite extension E/F in $\overline{\mathbf{Q}}$, where $\overline{[W_E \colon W_E]}$ is the closure of the commutator subgroup $[W_E : W_E]$. The system is characterized by the following properties:

(W1) The composite: $C_E \xrightarrow{r_E} W_E^{ab} \xrightarrow{\varphi} \mathrm{Gal}(E_{ab}/E)$ gives the Artin reciprocity map of class field theory for E;

(W2) The following diagrams are commutative: for $\sigma = \varphi(w) \in \mathrm{Gal}(\overline{\mathbf{Q}}/\mathbf{Q})$ and for E/F

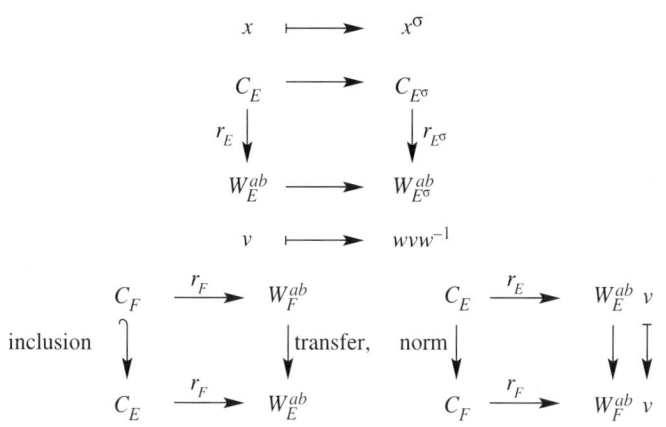

$$\begin{array}{ccc} x & \longmapsto & x^\sigma \\ C_E & \longrightarrow & C_{E^\sigma} \\ {\scriptstyle r_E}\downarrow & & \downarrow {\scriptstyle r_{E^\sigma}} \\ W_E^{ab} & \longrightarrow & W_{E^\sigma}^{ab} \\ v & \longmapsto & wvw^{-1} \end{array}$$

$$\begin{array}{ccccccc} & C_F & \xrightarrow{r_F} & W_F^{ab} & & C_E & \xrightarrow{r_E} & W_E^{ab}\,v \\ \text{inclusion} & \downarrow & & \downarrow \text{transfer,} & \text{norm} \downarrow & & \downarrow \downarrow \\ & C_E & \xrightarrow{r_F} & W_E^{ab} & & C_F & \xrightarrow{r_F} & W_F^{ab}\,v \end{array}$$

where the transfer map is defined as follows: For each group G and its subgroup, choosing a complete representative set R for $H\backslash G$, we define the transfer $t: G/[G:G] \to H/[H:H]$ by $t(g[G:G]) = \prod_{x \in R} h_x \mod [H:H]$;

(W3) Writing $W_{F/\mathbf{Q}} = W_Q/[W_F : W_F]$, we have that groups. $W_\mathbf{Q} = \varprojlim_F W_{F/\mathbf{Q}}$

as topological

The existence of the system $\{W_F\}$ is shown by using the data of the "canonical 2-cocycle" appearing in class field theory, and hence, is highly artificial (see [**T79**] for the construction). Similarly, we can also construct the local Weil group by replacing finite extensions $F/\mathbf{Q}$ (resp. C_F) by finite extensions $F/\mathbf{Q}_p$ (resp. $F^\times$) and by using local class field theory in place of global class field theory. In this local case, we can actually see the group in an explicit manner. Write k_E for the residue field of an algebraic extension $E/\mathbf{Q}_p$; in particular, $k_{\overline{\mathbf{Q}}_p} = \overline{k}$ is an algebraic closure of k_F. Then Frob: $x \mapsto x^q$ for $q = \#k_F$ is a topological generator of $\mathrm{Gal}(\overline{k}/k_F)$. Then W_F is canonically given by the subgroup of $\mathrm{Gal}(\overline{\mathbf{Q}}_p/F)$ consisting of automorphisms σ which induce on $\overline{k}$ a power of Frob. The kernel of the natural map: $W_F \to \mathrm{Gal}(\overline{k}/k_F)$ (with dense image) is called the inertia subgroup of W_F. We write I_p for the inertia group of $W_{\mathbf{Q}_p}$. Then we define $r_F: F^\times \to W_F^{ab}$ by the reciprocity map such that $r_F(a)$ induces $x \mapsto x^{|a|_p^{-d}}$ on $\overline{k}$ for $a \in F^\times$ and $d = [F : \mathbf{Q}_p]$. Then the compatibility of local and global class field theory implies the existence of a natural morphism $\theta_p: W_{\mathbf{Q}_p} \to W_\mathbf{Q}$ making the following diagram commutative:

$$\begin{array}{ccc} W_{\mathbf{Q}_p} & \xrightarrow{\mathrm{Inc}} & \mathrm{Gal}(\overline{\mathbf{Q}}_p/\mathbf{Q}_p) \\ {\scriptstyle \theta_p}\downarrow & & \cap \\ W_\mathbf{Q} & \xrightarrow{\varphi} & \mathrm{Gal}(\overline{\mathbf{Q}}/\mathbf{Q}). \end{array}$$

Of course, the last vertical inclusion map depends on our embedding of $\overline{\mathbf{Q}}$ into $\overline{\mathbf{Q}}_p$. We have a natural Frobenius element $\mathrm{Frob}_p = \theta_p(\mathrm{Frob})$ in $W_\mathbf{Q}$ modulo I_p. For any continuous representation $\xi\colon W_Q \to \mathrm{End}_{\mathbf{C}}(V)$ for a finite-dimensional **C**-vector space V, we consider the subspace V^{I_p} fixed by the inertia group I_p. Then $\xi(\mathrm{Frob}_p)$ has meaning on V^{I_p} because Frob is determined modulo I_p. Then we define the L-function of ξ by

$$L(s,\xi) = \prod_p \det((\mathrm{id} - p^{-s}\xi(\mathrm{Frob}_p))|_{V^{I_p}})^{-1}.$$

As A. Weil proved, these L-functions (which converge on a suitable right half-plane) have meromorphic continuations on the whole complex plane and satisfy a good functional equation. Our desire is to find a good classification map

$\iota\colon$ {isomorphism classes of continuous irreducible representations of

W_Q into $\mathrm{GL}_n(\mathbf{C})\} \to \left\{\text{irreducible representations occurring on } \bigoplus_\xi L_2^0(\xi)\right\}$

preserving L-functions (i.e., $L(s,\iota(\sigma)) = L(s,\sigma)$) and to describe its image. We can think of the corresponding problem for W_{Q_p}. This problem is called the local Langlands conjecture [**T79**, 4.1.7]), which has been proven for many values of n (see [**He85**] and [**Ku87**]), that is, supercuspidal representations of $\mathrm{GL}_n(\mathbf{Q}_p)$ are classified by representations of the local Weil group via L-functions and its ε-factors. Actually, enlarging the Weil group a little (the enlarged group is called the Deligne-Weil group W'_{Q_p}), we can extend this conjecture to a correspondence between all semisimple representations of W'_{Q_p} and local admissible representations of $\mathrm{GL}_n(\mathbf{Q}_p)$ (see [**T79**, §4]). The local conjecture is successfully solved because the local Galois group is solvable (that is, it can be approximated by abelian groups). The Weil group, constructed out of all known abelian data from class field theory, tries to describe nonabelian objects. Therefore, it is like peering into the whole nonabelian world through a small hole of the established abelian theory. Since the global Galois group is highly nonabelian, the image of ι is small [**T79**, (2.1–2.3)]. The representations of W_Q consist of representations of Artin type, i.e., those having finite image (factoring through $\mathrm{Gal}(\overline{\mathbf{Q}}/\mathbf{Q})$), representations of Hecke type, i.e., twists of representations of Artin type by Hecke characters, and induced representations of previous types. We call automorphic representations in the image representations of Weil type. Even the Hasse-Weil zeta function of modular elliptic curves without complex multiplication cannot be associated to an automorphic representation of Weil type. Thus, we need to look at all the algebraic objects, simultaneously, which yield "good" (abelian and nonabelian) Galois representations. This idea is first conceived by Grothendieck as his theory of motives and is later elaborated by Deligne [**DMOS82**]. Admitting the standard conjectures, we see that the category of motives is (Tannakian and) equivalent to the (Tannakian) category of algebraic representations of a huge proalgebraic group $\mathscr{G}$ which is an extension of $\mathrm{Gal}(\overline{\mathbf{Q}}/\mathbf{Q})$. All irreducible algebraic representations of the motivic Galois group $\mathscr{G}$ should be somehow classified by algebraic automorphic representations (Langlands' hypothesis), but even the formulation of this expectation is not yet clear (see [**C90**]). Here the word "algebraic" means, on the Galois side, a morphism of the proalgebraic group $\mathscr{G}$ into the algebraic group $\mathrm{GL}(n)$ and, on the automorphic side, that the finite part of the automorphic representation is defined over a number field.

Under the isomorphism $\alpha\colon \mathrm{Gal}(\mathbf{Q}_{\mathrm{ab}}/\mathbf{Q}) \cong \mathbf{A}^\times/\mathbf{Q}^\times \mathbf{R}_+^\times$, the group $\mathbf{Z}_p^\times$ corresponds to the image of the inertia group I_p at p in $\mathrm{Gal}(\overline{\mathbf{Q}}/\mathbf{Q})$. Thus, it is natural to call a Hecke character $\xi\colon \mathbf{Q}^\times\backslash\mathbf{A}^\times \to \mathbf{C}^\times$ unramified at p if ξ is trivial on $\mathbf{Z}_p^\times$. A Hecke character ξ is unramified at all but finitely many primes p. A Hecke character ξ is called algebraic if $\xi(z_\infty) = z_\infty^m$ for $z_\infty \in \mathbf{R}_+^\times$. This definition is compatible with the definition of algebraic automorphic representations, i.e., ξ is algebraic in the above sense if and only if $\xi(z)$ stays in a number field E for all $z \in \mathbf{A}_f^\times$. According to Weil, we can attach to an algebraic ξ a (continuous) Galois character $\xi_\mathscr{I}$ into $E_\mathscr{I}^\times$ for the $\mathscr{I}$-adic completion $E_\mathscr{I}$ of E for each prime ideal $\mathscr{I}$ of E. The character $\xi_\mathscr{I}$ is characterized by the following properties:

(i) If p is prime to $\mathscr{I}$ then $\xi_\mathscr{I}$ is unramified at p if and only if ξ is unramified at p;

(ii) If p is prime to $\mathscr{I}$ and ξ is unramified at p, then $\xi_\mathscr{I}(\mathrm{Frob}_p) = \xi(p_p)$, where p_p is the image of $p \in \mathbf{Q}_p^\times$ (but not the image of $p \in \mathbf{Q}^\times$) in $\mathbf{A}^\times$.

We then define
$$L(s,\xi) = \prod_{p \text{ unramified}} (1 - \xi(p_p)p^{-s})^{-1} = \prod_{p \text{ unramified}} (1 - \xi_\mathscr{I}(\mathrm{Frob}_p)p^{-s})^{-1}.$$

Thus, if we consider Galois characters having values in an $\mathscr{I}$-adic field, then we do not need to enlarge the Galois group. This fact is partially true in the nonabelian case. To each (Grothendieck) motive $M_{/\mathbf{Q}}$ of rank n with coefficients in a number field E, we can associate a compatible system $\xi = \{\xi_\mathscr{I}\}_\mathscr{I}$ of $\mathscr{I}$-adic Galois representations indexed by prime ideals of E. Without referring to motives, we can define a compatible system ξ of $\xi_\mathscr{I}\colon \mathrm{Gal}(\overline{\mathbf{Q}}/\mathbf{Q}) \to \mathrm{GL}_n(E_\mathscr{I})$ as follows. First of all, $\xi_\mathscr{I}$ is unramified outside Nl for an integer N independent of $\mathscr{I}$ (l is the residual characteristic of $\mathscr{I}$). Let $V(\xi_\mathscr{I})$ be the representation space of $\xi_\mathscr{I}$ and consider the subspace $V(\xi_\mathscr{I})^{I_p}$ fixed by I_p. Then $\xi(\mathrm{Frob}_p)$ actually acts on $V(\xi_\mathscr{I})^{I_p}$. Here the system $\xi = \{\xi_\mathscr{I}\}_\mathscr{I}$ is called compatible if
$$L_{\xi,p}(X) = \det((1 - X\xi_\mathscr{I}(\mathrm{Frob}_p))|_{V(\xi_\mathscr{I})^{I_p}}) \in E[X] \text{ for all } p \text{ outside } Nl$$
is independent of $\mathscr{I}$. We then define
$$L(s,\xi) = \prod_p L_{\xi,p}(p^{-s})^{-1}.$$

When ξ is attached to a motive M, this L-function is called the L-function of the motive M and is written as $L(s, M)$. If the classification map referred to as the Langlands hypothesis
$$\iota\colon \{\text{simple motives}\} \to \{\text{algebraic irreducible cuspidal automorphic representations}\}$$
exists, then it must satisfy the identity of L-functions: $L(s, \iota(M)) = L(s, M)$.

§6

We now turn to the p-adic case. Hereafter, unless otherwise mentioned, we suppose that $p > 2$ and $n = 2$ (i.e., we consider the algebraic group $\mathrm{GL}(2)_{/\mathbf{Q}}$). The fundamental question is:

What type of space should we take as a p-adic analog of $L_2^0(\xi)$?

Here we only think about automorphic representations of p-power conductors to make our argument simple. Such representations always possess a vector invariant under the group $U(p^\infty)$ of the matrices in $\mathrm{GL}_2(\widehat{\mathbf{Z}})$ which are unipotent at p. Thus, even in the complex case, we can replace $L_2^0(\xi)$ by invariant vectors under $U(p^\infty)$. There are several different choices of such analogs. A naive (algebro-geometric) definition of the space of p-adic modular forms is possible [**K78**], but here we give a cohomological definition of a p-adic analog of $L_2^0(\xi)$ which can be easily extended to $\mathrm{GL}(2)_{/F}$ for all number fields F. We consider the following open compact subgroups of $\mathrm{GL}_2(\mathbf{A}_f)$:

$$\text{(9)} \quad U(p^r) = \left\{ \begin{pmatrix} a & b \\ c & d \end{pmatrix} \in U_1(N) \middle| \begin{pmatrix} a & b \\ c & d \end{pmatrix} \equiv \begin{pmatrix} 1 & * \\ 0 & 1 \end{pmatrix} \mod p^r \widehat{\mathbf{Z}} \right\} \quad \text{and}$$

$$U(p^\infty) = \bigcap_r U(p^r).$$

The group $\mathrm{GL}_2(\mathbf{R})$ acts on $\mathbf{C} - \mathbf{R}$ via $\gamma(z) = \frac{az+b}{cz+d}$ for $\gamma = \begin{pmatrix} a & b \\ c & d \end{pmatrix}$. We then set $\mathrm{GL}_2^+(\mathbf{R}) = \{\gamma \in \mathrm{GL}_2(\mathbf{R}) \mid \gamma(H) = H\}$ for the upper half complex plane $H = \{z \in \mathbf{C} \mid \mathrm{Im}(z) > 0\}$. Then for $\mathrm{GL}_2^+(\mathbf{R}) = \{x \in \mathrm{GL}_2(\mathbf{R}) \mid \det(x) > 0\}$, we have $\mathrm{GL}_2^+(\mathbf{R})/C_\infty \cong H$ via $g \mapsto g(\sqrt{-1})$, where $C_\infty = \mathrm{SO}_2(\mathbf{R})\mathbf{R}^\times$ is the pull-back image of the standard maximal compact subgroup of $\mathrm{PGL}_2(\mathbf{R})$. For each open compact subgroup S of $\mathrm{GL}_2(\mathbf{A}_f)$, we consider the quotient topological space

$$X(S) = \mathrm{GL}_2(\mathbf{Q}) \backslash \mathrm{GL}_2(\mathbf{A}) / SC_\infty \cong \mathrm{GL}_2^+(\mathbf{Q}) \backslash \mathrm{GL}_2^+(\mathbf{A}) / SC_\infty,$$

where $\mathrm{GL}_2^+(\mathbf{A}) = \mathrm{GL}_2(\mathbf{A}_f) \times \mathrm{GL}_2^+(\mathbf{R})$ and $\mathrm{GL}_2^+(\mathbf{Q}) = \mathrm{GL}_2(\mathbf{Q}) \cap \mathrm{GL}_2^+(\mathbf{A})$. If S is sufficiently small so that $S \subset U(p^r)$ for some $p^r \geq 4$, then $X(S)$ is an open Riemann surface with finitely many connected components which are isomorphic to $\Gamma \backslash H$ for some congruence subgroup Γ ($\cong \mathrm{SL}_2(\mathbf{Z}) \cap S$) of $\mathrm{SL}_2(\mathbf{Z})$ ([**Sh71**, Chapter 6] and [**H93b**]). In particular, we write $X(p^r)$ for $X(U(p^r))$.

Let A be a complete valuation ring in $\overline{\mathbf{Q}}_p$ with finite residue field, and let K be its quotient field. Assume that $[K : \mathbf{Q}_p] < +\infty$. Take an A-module M with an action of

$$S_p^t = \{u_p^t = \det(u_p) u_p^{-1} \in \mathrm{GL}_2(\mathbf{Z}_p) \mid u \in S\}.$$

Then we let $\mathrm{GL}_2(\mathbf{Q})$ (resp., SC_∞) act on $\mathrm{GL}_2(\mathbf{A}) \times M$ from the left (resp., from the right) as follows: $\alpha(x,m)u = (\alpha x u, u_p^t m)$. This action is a discrete action if we use the discrete topology on M and the product topology on $\mathrm{GL}_2(\mathbf{A}) \times M$. We then define the quotient space $\underline{M} = \mathrm{GL}_2(\mathbf{Q}) \backslash \{\mathrm{GL}_2(\mathbf{A}) \times M\} / SC_\infty$. This space $\underline{M}$ is a covering space of the Riemann surface $X(S)$ which is locally homeomorphic to $X(S)$. We consider the sheaf of continuous sections of $\underline{M} = \underline{M}_S$ on $X(S)$. We also write $\underline{M}$ for this sheaf. Then we can consider the cohomology groups

$$H^1(X(S), \underline{M}), \quad H_c^1(X(S), \underline{M}), \quad \text{and} \quad H_P^1(X(S), \underline{M}),$$

where $H_c^1(X(S), \underline{M})$ is the usual cohomology group of compact support and $H_P^1(X(S), \underline{M})$ is the natural image of $H_c^1(X(S), \underline{M})$ in the usual cohomology group $H^1(X(S), \underline{M})$ (see [**H93b**, Chapter 6]). Then if $S \supset W$ are two open compact subgroups, writing H^1_* for any one of H^1, H^1_c, and H^1_P, we have the restriction map and the trace map [**H93b**, §6.3]:

$$\operatorname{res}_{S/V} \colon H^1_*(X(S), \underline{M}) \to H^1_*(X(W), \underline{M})$$

and

$$\operatorname{Tr}_{S/V} \colon H^1_*(X(W), \underline{M}) \to H^1_*(X(S), \underline{M}).$$

Take two open subgroups S and V. Let $W = \alpha S \alpha^{-1} \cap V$, and let $W^\alpha = \alpha^{-1} W \alpha = S \cap \alpha^{-1} V \alpha$ for $\alpha \in \operatorname{GL}_2(\mathbf{A}_f)$. Suppose that M is actually a module over the semigroup generated by α'_p, S'_p, and V'_p in $\operatorname{GL}_2(\mathbf{Q}_p)$. Then we can define a morphism of sheaves $[\alpha] \colon \underline{M}_W \to \underline{M}_{W^\alpha}$ by $[\alpha](x, m) = (x\alpha, \alpha'_p m)$. It is easy to see that this is well defined. Then by the functoriality of cohomology theory, $[\alpha]$ induces a morphism

$$[\alpha] \colon H^1_*(X(W), \underline{M}) \to H^1_*(X(W^\alpha), \underline{M}).$$

Then we can define the Hecke operator $[S\alpha V] \colon H^1_*(X(S), \underline{M}) \to H^1_*(X(V), \underline{M})$ by

$$[S\alpha V] = \operatorname{Tr}_{V/W^\alpha} \circ [\alpha] \circ \operatorname{res}_{S/W}.$$

It is easy to see that the operator $[S\alpha V]$ depends only on the double coset $S\alpha V$ but not on the choice of α. We now apply the above argument to the following modules. Let $\xi \colon (\mathbf{Z}/p^r\mathbf{Z})^\times \to A^\times$ be a character. Let B be any A-module. Let $M = L(n, v, \xi; B)$ for $(n, v) \in \mathbf{Z}^2$ ($n \geq 0$) be the A-module of homogeneous polynomials in $B[X, Y]$ of degree n. Here $B[X, Y]$ denotes the polynomial module (in X and Y) with coefficients in B. We let $S(p^r)'_p$ act on $P(X, Y)$ by

(S) $$uP(X, Y) = \xi(d) \det(u)^v P((X, Y)^t u^t) \quad \text{for } u = \begin{pmatrix} a & b \\ c & d \end{pmatrix}.$$

Then we write $\mathscr{L}(n, v, \xi; B)$ for $\underline{M}$. For the semigroup

(Δ) $$\Delta = \left\{ \alpha = \begin{pmatrix} a & b \\ c & d \end{pmatrix} \in \operatorname{GL}_2(\mathbf{A}_f) \mid \alpha_p \in M_2(\mathbf{Z}_p), a_p \in \mathbf{Z}_p^\times, c \in p^r \mathbf{Z}_p \right\},$$

we let Δ' act on $L(n, v, \xi; A)$ by setting

$$\alpha^t P(X, Y) = \xi(a_p) \det(\alpha)^v P((X, Y)^t \alpha_p).$$

Thus, we have the operator $[S\alpha S]$ for $\alpha \in \Delta$. To the triple (n, v, ξ), we attach a character of the group $\mathbf{Z}_p^\times \times \mathbf{Z}_p^\times$ given by $(w, z) \mapsto w^v z^{n+2v} \xi(z)$. Then it is known that $H^1_P(X(p^s), \mathscr{L}(n, v, \xi; K)) = 0$ if the character attached to (n, v, ξ) does not factor through $\mathbf{G} = \mathbf{Z}_p^\times \times (\mathbf{A}^\times / \mathbf{Q}^\times U^{(p)} \mathbf{R}^\times)$, where we consider $\mathbf{G}$ as a quotient of $\mathbf{Z}_p^\times \times \mathbf{Z}_p^\times$ via the isomorphism $\chi \colon \mathbf{A}^\times / \mathbf{Q}^\times U^{(p)} \mathbf{R}^\times \cong \mathbf{Z}_p^\times / \{\pm 1\}$ (see (7)). We call a character φ of $\mathbf{G}$ arithmetic if there exist integers $n \geq 0$ and v such that in a small neighborhood of 1 in $\mathbf{G}$, φ coincides with the character: $(w, z) \mapsto w^v z^{n+2v}$.

An arithmetic character φ of $\mathbf{G}$ determines the data (n, v, ξ). Thus, we shall hereafter write $\mathscr{L}(\varphi; B)$ for $\mathscr{L}(n, v, \xi; B)$. Using this action, we define the operator

$$T(z) = z_p^{-v} \left[S \begin{pmatrix} 1 & 0 \\ 0 & z \end{pmatrix} S \right] \quad (S = U(p^r)) (z \in \mathscr{Z} = \mathbf{A}_f^\times \cap \widehat{\mathbf{Z}}).$$

There is another set of operators acting on $H^1_P(X(p^r), \mathscr{L}(\varphi; B))$ for $X(p^r) = X(U(p^r))$. For each $z \in \mathscr{Z}$ with $z_p \in \mathbf{Z}_p^\times$, we can think of the action $\langle z \rangle = \varphi(1, z_p)^{-1} [SzS]$. We have written $T(q)$ (resp., $\langle q \rangle$) as $T_1(q)$ (resp., $T_2(q)$) for primes

q prime to p in paragraph 4 (see (L)). These operators $\{T(n), \langle z \rangle\}$ are mutually commutative and compatible with the restriction map

$$\operatorname{res}_{r,s}: H^1_P(X(p^r), \mathscr{L}(\varphi; B)) \to H^1_P(X(p^s), \mathscr{L}(\varphi; B)) \quad \text{for } s \geq r.$$

Therefore, we have

$$T(z) \circ \operatorname{res}_{r,s} = \operatorname{res}_{r,s} \circ T(z) \quad \text{and} \quad \langle z \rangle \circ \operatorname{res}_{r,s} = \operatorname{res}_{r,s} \circ \langle z \rangle.$$

To define a Hecke algebra, we can take as B either K or K/A. The two choices of B yield the same result. We define the Hecke algebra $\mathbf{h}_\varphi(p^r; A)$ to be the A-subalgebra of the A-linear endomorphism algebra of $H^1_P(X(p^r), \mathscr{L}(\varphi; B))$ ($B = K$ or K/A) generated by $T(z)$ and $\langle z \rangle$ for all $z \in \mathscr{Z}$. The algebra $\mathbf{h}_\varphi(p^r; A)$ is commutative with identity $T(1)$ and is free of finite rank over A. Therefore, it is a compact ring. The restriction of these operators acting on $H^1_P(X(p^s), \mathscr{L}(\varphi; K))$ to $H^1_P(X(p^r), \mathscr{L}(\varphi; K))$ yields an A-algebra homomorphism of $\mathbf{h}_\varphi(p^s; A)$ onto $\mathbf{h}_\varphi(p^r; A)$ for $s > r$. We then take the following two limits

$$\mathscr{V}(B)_{/\mathbf{Q}} = H^1_P(X(p^\infty), \mathscr{L}(\varphi; B)) = \varinjlim_n H^1_P(X(p^n), \mathscr{L}(\varphi; B))$$

and

$$\mathbf{h}_\varphi(p^\infty; A) = \varprojlim_n \mathbf{h}_\varphi(p^n; A).$$

Then the Hecke algebra $\mathbf{h}_\varphi(p^\infty; A)$ is a compact ring and naturally acts on $\mathscr{V}(B)$. In fact, we can define the p-adic topology on $\mathscr{V}(K)$ so that the natural images of $\{p^n \mathscr{V}(A)\}$ form a fundamental system of neighborhoods of 0. Then the Hecke algebra acts on $\mathscr{V}(K)$ via bounded operators. We may take the p-adic completion $\overline{\mathscr{V}}(K)$ of $\mathscr{V}(K)$ as a p-adic analog of the subspace of $L^0_2(\xi)$ (or even the symbolic continuous sum $\int L^0_2(\xi)\,d\xi$) fixed by $U(p^\infty) = \bigcap_r U(p^r)$. The important fact is that the pair

$$(\mathbf{h}_\varphi(p^\infty; A), \{T(z), \langle z \rangle\})$$

is independent of φ. Hence, there exists an isomorphism of $\mathbf{h}_\varphi(p^\infty; A)$ onto $\mathbf{h}_\psi(p^\infty; A)$ for any two arithmetic characters φ and ψ of $\mathbf{G}$ which takes $T(z)$ to $T(z)$ and $\langle z \rangle$ to $\langle z \rangle$ [H89b]. Thus, we may write this universal object as $\mathbf{h}_{/\mathbf{Q}} = \mathbf{h}(p^\infty; A)_{/\mathbf{Q}}$. Then the algebro-geometric spectrum $\operatorname{Spec}(\mathbf{h})$ is a p-adic analog of the complex spectrum of $\int L^0_2(\xi)\,d\xi$. It seems that in the p-adic case the spectrum is large compared to the complex case. We have two continuous characters

$$\langle\,\rangle: \mathbf{A}^\times/U^{(p)}\mathbf{Q}^\times\mathbf{R}^\times = \mathbf{Z}_p^\times/\{\pm 1\} \to \mathbf{h}_{/\mathbf{Q}}: z \mapsto \langle z \rangle$$

and

$$T: \mathbf{Z}_p^\times \to \mathbf{h}_{/\mathbf{Q}}: u \mapsto T(u).$$

This extends to an algebra homomorphism

(10) $\quad T \times \langle\,\rangle: \mathbf{Z}_p[[G]] \to \mathbf{h}_{/\mathbf{Q}} \quad (\text{and } T \times \langle\,\rangle: \mathbf{Z}_p[[W \times W]] \to \mathbf{h}_{/\mathbf{Q}}).$

We thus propose to take the space $\operatorname{Spec}(\mathbf{h})(\overline{\mathbf{Q}}_p) = \operatorname{Hom}_{A\text{-alg}}(\mathbf{h}, \overline{\mathbf{Q}}_p)$ (or its irreducible components) as a natural domain on which modular p-adic L-functions should be defined. As was already seen in §3, the connected and irreducible component of $\operatorname{Spec}(\mathbf{Z}_p[[G]])$ is given by $\operatorname{Spec}(\mathbf{Z}_p[[W]])$, and $\operatorname{Spec}(\mathbf{Z}_p[[W]])(\overline{\mathbf{Q}}_p)$ is isomorphic to the p-adic open unit disk D (see (8)). Since the morphism (10) induces a covering map $\text{Spec}(\mathbf{h})(\overline{\mathbf{Q}}_p) \to D \times D$, we are tempted to conjecture

CONJECTURE. *The dimension of each irreducible component of* $\text{Spec}(\mathbf{h})(\overline{\mathbf{Q}}_p)$ *as a p-adic space is greater than or equal to 2. Moreover, the natural covering* $\text{Spec}(\mathbf{h}) \to \text{Spec}(\mathbf{Z}_p[[W \times W]])$ *is dominant on each irreducible component.*

Here we implicitly assumed that $\mathbf{h}$ is Noetherian. We will discuss this question later (Corollary 3). The conjecture also implies that the Krull dimension of the coordinate ring of each irreducible component is greater than or equal to 3. Here for the first time we encounter the situation where the dimension of the space on which a p-adic L-function should be defined is larger than 1. We will later see that some irreducible components are finite coverings of $D \times D$, and therefore, have dimension 2 (Theorem 6 and [**H93b**], §7.6]).

We can almost automatically extend the above definition of $\mathbf{h}$ to any number field F. We briefly explain how to define the p-adic Hecke algebra for $\text{GL}(2)_{/F}$. Since $\mathbf{A} = \mathbf{A}_f \times \mathbf{R}$ for $\mathbf{A}_f = \{z \in \mathbf{A} \mid z_\infty = 0\}$, we have $F_\mathbf{A} = F_{\mathbf{A}_f} \times F_\mathbf{R}$ for $F_{\mathbf{A}_f} = F \otimes_\mathbf{Q} \mathbf{A}_f$ and $F_\mathbf{R} = F \otimes_\mathbf{Q} \mathbf{R}$. Let I be the set of all fields embedding F into $\overline{\mathbf{Q}}$ and fix two embeddings $i_\infty: \overline{\mathbf{Q}} \to \mathbf{C}$ and $i_p: \overline{\mathbf{Q}} \to \overline{\mathbf{Q}}_p$. Let A and K be as above, but we assume that K contains F^σ for all $\sigma \in I$. Let O be the integer ring of F, and set $\mathscr{O}_p = \mathscr{O} \otimes_\mathbf{Z} \mathbf{Z}_p$ and $\widehat{\mathscr{O}} = \mathscr{O} \otimes_\mathbf{Z} \widehat{\mathbf{Z}}$, which will be regarded as a subring of $F_{\mathbf{A}_f}$. Then we define subgroups $U(p^r)$ of $\text{GL}_2(F_{\mathbf{A}_f})$ in the same way as (9) replacing $\widehat{\mathbf{Z}}$ by $\widehat{\mathscr{O}}$. Similarly, we write C_∞ for the pull-back image of the standard maximal compact subgroup of $\text{PGL}_2(F_\mathbf{R})$ in the identity (connected) component $\text{GL}_2^+(F_\mathbf{R})$. We define a topological group $\mathbf{G}$ and G by

$$\mathbf{G} = \mathscr{O}_p^\times \times (F_\mathbf{A}^\times / \overline{F^\times U^{(p)} F_\mathbf{R}^\times}) \supset \mathscr{O}_p^\times \times (\mathscr{O}_p^\times / \overline{\mathscr{O}^\times}) = G,$$

where $U^{(p)} = \{z \in \widehat{\mathscr{O}}^\times \mid z_p = 1\}$. It is easy to see that G is an open compact subgroup of $\mathbf{G}$. A character $\varphi: \mathbf{G} \to K^\times$ is called arithmetic if there exist I-tuples of integers $n = (n_\sigma \geq 0)_{\sigma \in I}$ and $v = (v_\sigma)_{\sigma \in I}$ such that on a small open neighborhood of the identity of G, φ coincides with $(w, z) \mapsto w^v z^{n+2v}$, where $w^v = \prod_{\sigma \in I} w^{\sigma v_\sigma}$. From an arithmetic φ, we recover the data (n, v, ξ) (which we write as $(n(\varphi), v(\varphi), \xi(\varphi))$ if necessary), where ξ is a finite-order character of $\mathscr{O}_p^\times$ given by $\xi(z) = \varphi(1, z) z^{-n-2v}$. For any A-module B, we consider the polynomial module $B[X_\sigma, Y_\sigma]_{\sigma \in I}$ with indeterminate $(X_\sigma, Y_\sigma)_{\sigma \in I}$. We write $L(\varphi; B)$ for the space of polynomials homogeneous in each pair (X_σ, Y_σ) of degree n_σ for the I-tuple $n = n(\varphi)$. If ξ factors through $(\mathscr{O}/p^r\mathscr{O})^\times$, then we get $U(p^r)_p^t$ act on $L(\varphi; B)$ by

(U) $\qquad uP(X_\sigma, Y_\sigma) = \xi(d) \det(u)^v P((X_\sigma, Y_\sigma)^t u^t) \quad \text{for } u = \begin{pmatrix} a & b \\ c & d \end{pmatrix}.$

For the semigroup

(Δ) $\qquad \Delta = \left\{ \alpha = \begin{pmatrix} a & b \\ c & d \end{pmatrix} \in \text{GL}_2(F_{\mathbf{A}_f}) \mid \alpha_p \in M_2(O_p), a_p \in \mathscr{O}_p^\times, c \in p^r \mathscr{O}_p \right\},$

we let Δ^t act on $L(\varphi; B)$ by

$$\alpha^t P(X_\sigma, Y_\sigma) = \xi(a_p) \det(\alpha_p)^v P((X_\sigma, Y_\sigma)^t \alpha_p^\sigma).$$

Then for each r, we can define a sheaf $\mathscr{L}(\varphi; B)$ of locally constant sections of the covering space
$$\mathrm{GL}_2(F)\backslash(\mathrm{GL}_2(F_A) \times L(\varphi; B))/U(p^r)C_\infty$$
over
$$X(p^r) = \mathrm{GL}_2(F)\backslash \mathrm{GL}_2(F_A)/U(p^r)C_\infty.$$
The topological space $X(p^r)$ is a Riemannian manifold of (real) dimension $2r_1 + 3r_2$ if r is sufficiently large, where r_1 (resp., r_2) is the number of real (resp., complex) places of F. We then consider $H_P^j(X(p^r), \mathscr{L}(\varphi; B))$. We can define Hecke operators acting on the cohomology group for $S = U(p^r)$ by
$$T(z) = z_p^{-v}\left[S\begin{pmatrix}1 & 0\\ 0 & z\end{pmatrix}S\right] \quad \text{for } z \in \mathscr{Z} = \widehat{\mathscr{O}} \cap F_{A_f}^\times$$
and
$$\langle z \rangle = \varphi(1, z_p)^{-1}[SzS] \quad \text{for } z \in \mathscr{Z} \text{ with } z_p \in \mathscr{O}_p^\times$$
in the same manner as in the case of $F = \mathbf{Q}$. For each prime ideal $\mathscr{I}$ of $\mathscr{O}$ prime to p, we write $T(\mathscr{I})$ (resp., $\langle \mathscr{I} \rangle$) for $T(\varpi)$ for a prime element ϖ of the $\mathscr{I}$-adic completion $\mathscr{O}_\mathscr{I}$ regarded as an element of F_A. The operators $T(\mathscr{I})$ and $\langle \mathscr{I} \rangle$ are well defined independently of the choice of ϖ. We know (see [**Ha87**] and [**H93a**]) that
$$H_P^j(X(S), \mathscr{L}(\varphi; K)) = \{0\} \quad \text{if } n(\varphi) \neq 0 \text{ and } j \notin [r_1 + r_2, r_1 + 2r_2],$$
and for all $j \in [r_1 + r_2, r_1 + 2r_2]$, the A-subalgebras $h_\varphi(p^r; A)_{/F}$ of
$$\mathrm{End}_K(H_P^j(X(p^r), \mathscr{L}(\varphi; K)))$$
generated by all $T(z)$ and $\langle z \rangle$ are canonically isomorphic to each other. Thus, we may concentrate our attention on $H_P^q(X(S), \mathscr{L}(\varphi; B))$ for $q = r_1 + r_2$ when B is a field of characteristic 0. We may also define $\mathbf{h}_\varphi(p^r; A)$ by the A-subalgebra of $\mathrm{End}_K(H_P^q(X(p^r), \mathscr{L}(\varphi; K/A)))$ generated by all $T(z)$ and $\langle z \rangle$. For this algebra, we actually need to consider H_P^j for general j, but for simplicity, we discuss only the qth cohomology group. Since we have a natural map of sheaves $\pi: \mathscr{L}(\varphi; K) \to \mathscr{L}(\varphi; K/A)$, we have an A-algebra homomorphism
$$\rho: \mathbf{h}_\varphi(p^r; A) \to h_\varphi(p^r; A) \quad \text{determined by } h \circ \pi = \pi \circ \rho(h).$$
Naturally ρ takes $T(z)$ and $\langle z \rangle$ to $T(z)$ and $\langle z \rangle$, and hence, ρ is surjective. In the same manner as in the case of $F = \mathbf{Q}$, we have two natural projective systems of triples: $\{h_\varphi(p^r; A), T(z), \langle z \rangle\}_r$ and $\{\mathbf{h}_\varphi(p^r; A), T(z), \langle z \rangle\}_r$. We form the following two limits:
$$\{h_\varphi(p^\infty; A), T(z), \langle z \rangle\} = \varprojlim_n \{h_\varphi(p^n; A), T(z), \langle z \rangle\}$$
and
$$\{\mathbf{h}_\varphi(p^\infty; A), T(z), \langle z \rangle\} = \varprojlim_n \{\mathbf{h}_\varphi(p^n; A), T(z), \langle z \rangle\}.$$
We then ask

QUESTION 1. Is ρ an isomorphism for $r = 1, 2, \ldots, \infty$?

It is known that ρ is an isomorphism when $F = \mathbf{Q}$ because $H^1(X(p^r), K/A)$ is divisible. When F is totally real, it is expected that $\mathrm{Ker}(\rho)$ is small (for example, is torsion or even pseudonull over $A[[\mathbf{G}]]$), but the pseudonullity is not yet proven. If F has complex places, the answer to Question 1 is known to be negative in some cases. Writing c for complex conjugation, it is known [**Ha87**] that $H_P^j(X(S), \mathscr{L}(\varphi; K)) = \{0\}$ if $n_\sigma \neq n_{\sigma c}$ for some $\sigma \in I$ (see also [**H93a**]). However, there are plenty of examples of nonvanishing of $H_P^j(X(S), \mathscr{L}(\varphi; K/A))$ even if $n \neq 0$ ([**Ta**], [**H93a**]).

QUESTION 2. Are the algebras $\mathbf{h}_\varphi(p^\infty; A)$ and $h_\varphi(p^\infty, A)$ independent of φ if $n(\varphi) \neq 0$?

When F is totally real, it is seen in [**H89b**] that $h_\varphi(p^\infty; A)$ is independent of φ. There is a partial result of this type for the nearly ordinary part of the ring $\mathbf{h}_\varphi(p^\infty; A)$ valid for arbitrary F (Theorem 4).

QUESTION 3. How many connected components does $\mathrm{Spec}(\mathbf{h}_\varphi(p^\infty; A))$ have?

This can be interpreted as asking: how many congruence classes modulo p are there of the eigenvalues of Hecke operators of p-power level? It is classically known [**Se75**] that there are only finitely many connected components when $F = \mathbf{Q}$, and it is not hard to extend the proof of the finiteness in [**Se75**] to totally real F, because we can now attach a Galois representation to such a system of eigenvalues ([**Ta89**], [**BR89**]).

We decompose $\mathbf{G} = \mu \times \mathbf{W}$ for a finite group μ and $\mathbf{W}$ isomorphic to $\mathbf{Z}_p^d$ for some $d = d(F) > 0$. We always have $d(F) = [F : \mathbf{Q}] + r_2 + 1 + \delta(p, F)$ for $\delta(p, F) \geq 0$ and $\delta(p, F) = 0$ if the Leopoldt conjecture is true for p and F. Each irreducible component of $\mathrm{Spec}(A[[\mathbf{G}]])$ is isomorphic to $\mathrm{Spec}(A[[\mathbf{W}]])$ whose space of $\overline{\mathbf{Q}}_p$-valued points is isomorphic to the product of d copies of the open unit disk: $D \times \cdots \times D$. In particular, the (Krull) dimension of $A[[\mathbf{W}]]$ is $d + 1$. We have a natural algebra homomorphism

$$T \times \langle \ \rangle \colon A[[\mathbf{G}]] \to \mathbf{h}_\varphi(p^\infty; A) \quad \text{and} \quad T \times \langle \ \rangle \colon A[[\mathbf{G}]] \to h_\varphi(p^\infty; A).$$

Thus, $\mathrm{Spec}(\mathbf{h}_\varphi(p^\infty; A))$ and $\mathrm{Spec}(h_\varphi(p^\infty; A))$ are covering spaces of $\mathrm{Spec}(A[[\mathbf{W}]])$.

QUESTION 4. Are there some irreducible components of $\mathrm{Spec}(h_\varphi(p^\infty; A))$ or $\mathrm{Spec}(\mathbf{h}_\varphi(p^\infty; A))$ that dominate $\mathrm{Spec}(A[[\mathbf{W}]])$?

When F is totally real, there are such components: the irreducible components of the nearly ordinary part (Theorem 6). When F has some complex places, the answer is probably negative (see Theorem 6 and [**H93a**, Theorem 5.2]).

§7

We now speculate why we expect that $\mathbf{h}_\varphi(p^\infty; A)_{/F}$ is independent of φ (as long as $n(\varphi) \neq 0$), and at the same time we explain how to construct the maximal GL(2)-extension $F_{\mathrm{GL}(2),p}$ unramified outside p which is the GL(2) analog of $\mathbf{Q}_{\mathrm{GL}(1),p} = \mathbf{Q}(1^{1/p^\infty})$. Let $h_\varphi^{(p)}(p^r; A)_{/F}(r = 1, 2, \ldots, \infty)$ be the closed subalgebra of $h_\varphi(p^r; A)_{/F}$ generated over $A[[\mathbf{W}]]$ by $T(z)$ with $z_p = 1$. It is well known from the theory of primitive (or new) forms (see [**Mi89**, Chapter 4] for $F = \mathbf{Q}$) that $h_\varphi^{(p)}(p^r; A)$ is reduced (i.e., has no nontrivial nilpotent elements). Since $h_\varphi^{(p)}(p^\infty; A)$ is the projective limit of $\{h_\varphi^{(p)}(p^r; A)\}_r$, $h^{(p)} = h_\varphi^{(p)}(p^\infty; A)$ is also reduced. For any profinite local algebra R over $A[[\mathbf{W}]]$ and an $A[[\mathbf{W}]]$-algebra homomorphism $\lambda \colon h_\varphi^{(p)}(p^\infty; A)_{/F} \to R$, a continuous Galois representation $\pi\colon \mathrm{Gal}(\overline{\mathbf{Q}}/F) \to \mathrm{GL}_2(R)$ is called λ-residual if the following conditions are satisfied:

(R1) π is unramified outside p;

(R2) for the Frobenius element $\mathrm{Frob}_{\mathscr{I}}$ for each prime ideal $\mathscr{I}$ prime to p, we have

$$\det(1_2 - \pi(\mathrm{Frob}_{\mathscr{I}})X) = 1 - \lambda(T(\mathscr{I}))X + N(\mathscr{I})\lambda(\langle\mathscr{I}\rangle)X^2.$$

The existence of a λ-residual representation is known, when F is totally real, in the following cases ([**M89**], [**Gv88**], [**Wi88**], [**Ta89**], [**BR89**], [**H89c**], [**HT93b**], [**H93b**, §7.5]):

(i) R is a field;

(ii) R is an integral domain and for the maximal ideal $\mathfrak{m}$ of R, the $(\lambda \bmod \mathfrak{m})$-residual representation is irreducible (see Corollary 1 below);

(iii) the $(\lambda \bmod \mathfrak{m})$-residual representation is absolutely irreducible.

Thus, the points of $\mathrm{Spec}(h^{(p)})$ parametrize residual representations. However, there may be several λ-residual representations for a given point λ. To get objects that are parametrized exactly by points of $\mathrm{Spec}(h^{(p)})$, we introduce the notion of pseudorepresentations first given by Wiles [**Wi88**]. Let $G = G_F$ be the Galois group of the maximal extension of F unramified outside p. Let π be a (continuous) representation of G into $\mathrm{GL}_2(R)$. We assume the existence of an element $c \in G$ of order 2 such that $\det(\pi(c)) = -1$. If F has a real place v, the complex conjugation at v is often taken as c. Then by the assumption that $p > 2$, we may assume that

$$\pi(c) = \begin{pmatrix} -1 & 0 \\ 0 & 1 \end{pmatrix}.$$

For each $\sigma \in G$, we write $\pi(\sigma) = \begin{pmatrix} a(\sigma) & b(\sigma) \\ c(\sigma) & d(\sigma) \end{pmatrix}$ and define a function $x\colon G \times G \to R$ by $x(\sigma, \tau) = b(\sigma)c(\tau)$. Then these functions satisfy the following properties:

(PR1) As functions on G or G^2, a, d, and x are continuous,

(PR2) $a(\sigma\tau) = a(\sigma)a(\tau) + x(\sigma, \tau)$, $d(\sigma\tau) = d(\sigma)d(\tau) + x(\tau, \sigma)$, and

$$x(\sigma\tau, \rho\gamma) = a(\sigma)a(\gamma)x(\tau, \rho) + a(\gamma)d(\tau)x(\sigma, \rho)$$
$$+ a(\sigma)d(\rho)x(\tau, \gamma) + d(\tau)d(\rho)x(\sigma, \gamma),$$

(PR3) $a(1) = d(1) = d(c) = 1$, $a(c) = -1$, and

$$x(\sigma, \rho) = x(\rho, \tau) = 0 \text{ if } \rho = 1 \text{ or } c,$$

(PR4) $x(\sigma, \tau)x(\rho, \eta) = x(\sigma, \eta)x(\rho, \tau)$.

The properties (PR3) and (PR4) follow directly from the definition, and the first half of (PR2) can be proven by computing directly the multiplicative formula

$$\begin{pmatrix} a(\sigma) & b(\sigma) \\ c(\sigma) & d(\sigma) \end{pmatrix} \begin{pmatrix} a(\tau) & b(\tau) \\ c(\tau) & d(\tau) \end{pmatrix} = \begin{pmatrix} a(\sigma\tau) & b(\sigma\tau) \\ c(\sigma\tau) & d(\sigma\tau) \end{pmatrix}.$$

Then in addition to the first two formulas of (PR2), we also have

$$b(\sigma\tau) = a(\sigma)b(\tau) + b(\sigma)d(\tau) \quad \text{and} \quad c(\sigma\tau) = c(\sigma)a(\tau) + d(\sigma)c(\tau).$$

Thus, we know that

$$x(\sigma\tau, \rho\gamma) = b(\sigma\tau)c(\rho\gamma) = (a(\sigma)b(\tau) + b(\sigma)d(\tau))(c(\rho)a(\gamma) + d(\rho)c(\gamma))$$
$$= a(\sigma)a(\gamma)x(\tau,\rho) + a(\gamma)d(\tau)x(\sigma,\rho) + a(\sigma)d(\rho)x(\tau,\gamma)$$
$$+ d(\tau)d(\rho)x(\sigma,\gamma).$$

For any topological algebra R, we now define a *pseudorepresentation* of G into R to be a triple $\pi' = (a, d, x)$ consisting of continuous functions on G and G^2 satisfying the conditions (PR1–4). We define the trace $\mathrm{Tr}(\pi')$ (resp., the determinant $\det(\pi')$) of the pseudorepresentation π' to be a function on G given by

$$\mathrm{Tr}(\pi')(\sigma) = a(\sigma) + d(\sigma) \quad (\text{resp.}, \ \det(\pi')(\sigma) = a(\sigma)d(\sigma) - x(\sigma, \sigma)).$$

Then we have

PROPOSITION 1 (Wiles). *Let $\pi' = (a, d, x)$ be a pseudorepresentation of G into an integral domain R with quotient field Q. Then there exists a continuous representation $\pi: G \to \mathrm{GL}_2(Q)$ with the same trace and determinant as π'.*

Here we only point out how to construct the representation π out of π' (see **[Wi88]**, **[H93b**, §7.5] for a detailed proof). We divide our argument into two cases:

CASE 1. there exist ρ and $\gamma \in G$ such that $x(\rho, \gamma) \neq 0$,
and

CASE 2. $x(\sigma, \tau) = 0$ for all σ, τ in G.

CASE 1. We define $\pi(\sigma) = \begin{pmatrix} a(\sigma) & b(\sigma) \\ c(\sigma) & d(\sigma) \end{pmatrix}$ by setting

$$c(\sigma) = x(\rho, \sigma) \quad \text{and} \quad b(\sigma) = x(\sigma, \gamma)/x(\rho, \gamma).$$

Then it is easy to check using (PR2, 4) that π actually gives a representation.

CASE 2. In this case, by (PR2) we have $a(\sigma)a(\tau) = a(\sigma\tau)$ and $d(\sigma)d(\tau) = d(\sigma\tau)$ for all $\sigma, \tau \in G$.

Then we simply put $\pi(\sigma) = \begin{pmatrix} a(\sigma) & 0 \\ 0 & d(\sigma) \end{pmatrix}$ which does the job.

The following corollary is obvious from the above explanation.

COROLLARY 1. *Let R be a profinite local $\mathbf{Z}_p$-algebra with maximal ideal $\mathfrak{m}$. Let $\pi' = (a, d, x)$ be a pseudorepresentation of G into R. Suppose there exist $\rho, \gamma \in G$ such that $x(\rho, \gamma) \in R^\times$. Then there exists a continuous representation $\pi: G \to \mathrm{GL}_2(R)$ with the same trace and determinant as π'. In particular, if the Galois representation into $\mathrm{GL}_2(R/\mathfrak{m})$ associated to π' mod $\mathfrak{m}$ exists and is irreducible, then we have a continuous representation $\pi: G \to \mathrm{GL}_2(R)$ with the same trace and determinant as π'.*

We write κ for the residue field of A. We consider the category $\mathscr{C}$ consisting of all profinite local A-algebras with residue field κ. For two maps $\alpha: X \to Z$ and $\beta: Y \to Z$ of sets, their fiber product is defined by $X \times_Z Y = \{(x, y) \in X \times Y \mid \alpha(x) = \beta(y)\}$. For three objects R, R_1, and R_2 in $\mathscr{C}$ and morphisms $\alpha_i: R_i \to R$, there exists a fiber product $R_1 \times_R R_2$ in $\mathscr{C}$ that is given by the set-theoretic fiber product. We write $\rho_i: R_1 \times_R R_2 \to R_i$ for the projection map. Let $\mathscr{PR}(R)$ be the set of all pseudorepresentations (with respect to c) of G_F having values in the ring R. Then we have a natural map

(11)
$$\gamma\colon \mathscr{PR}(R_1 \times_R R_2) \to \mathscr{PR}(R_1) \times_{\mathscr{PR}(R)} \mathscr{PR}(R_2) \quad \text{given by } \gamma(\pi) = \rho_1 \circ \pi \times \rho_2 \circ \pi.$$

PROPOSITION 2.

(i) *The canonical morphism* $\gamma\colon \mathscr{PR}(R_1 \times_R R_2) \to \mathscr{PR}(R_1) \times_{\mathscr{PR}(R)} \mathscr{PR}(R_2)$ *is a bijection.*

(ii) (Wiles) *Let a and ℓ be two ideals of R. Let $\pi(a)$ and $\pi(\ell)$ be pseudorepresentations into R/a and R/ℓ, respectively. Suppose that $\pi(a)$ and $\pi(\ell)$ are compatible; that is, there exist functions* tr *and* det *on a dense subset Σ of G with values in $R/a \cap \ell$ such that for all $\sigma \in \Sigma$,*

$$\operatorname{Tr}(\pi(a)(\sigma)) \equiv \operatorname{tr}(\sigma) \bmod a \quad \text{and} \quad \operatorname{Tr}(\pi(\ell)(\sigma)) \equiv \operatorname{tr}(\sigma) \bmod \ell,$$
$$\det(\pi(a)(\sigma)) \equiv \det(\sigma) \bmod a \quad \text{and} \quad \det(\pi(\ell)(\sigma)) \equiv \det(\sigma) \bmod \ell.$$

Then there exists a pseudorepresentation $\pi(a \cap \ell)$ of G into $R/a \cap \ell$ such that

$$\operatorname{Tr}(\pi(a \cap \ell)(\sigma)) = \operatorname{tr}(\sigma) \quad \text{and} \quad \det(\pi(a \cap \ell)(\sigma)) = \det(\sigma) \quad \text{on } \Sigma.$$

PROOF. The first assertion is obvious from the definition. To prove (ii), we consider the exact sequence

$$0 \to R/a \cap \ell \to R/a \oplus R/\ell \xrightarrow{\alpha} R/a + \ell \to 0,$$
$$a \mapsto a \bmod a \oplus a \bmod \ell, \quad a \oplus b \mapsto (a - b) \bmod (a + \ell).$$

We consider the pseudorepresentation $\pi = \pi(a) \oplus \pi(\ell)$ with values in $R/a \oplus R/\ell$. The function $\alpha \circ \operatorname{Tr}(\pi)$ vanishes identically on Σ. Since this function is continuous on G and Σ is dense in G, $\alpha \circ \operatorname{Tr}(\pi)$ vanishes on G. Thus, $\operatorname{Tr}(\pi)$ has values in $R/a \cap \ell$. If we write $\pi = (a, d, x)$, then

$$a(\sigma) = 2^{-1}(\operatorname{Tr}(\pi(\sigma)) - \operatorname{Tr}(\pi(\sigma c))), \quad d(\sigma) = 2^{-1}(\operatorname{Tr}(\pi(\sigma)) + \operatorname{Tr}(\pi(\sigma c))),$$

and

$$x(\sigma, \tau) = a(\sigma\tau) - a(\sigma)a(\tau).$$

Thus π itself has values in $R/a \cap \ell$ and gives the desired pseudorepresentation. □

A point $\lambda\colon h_\varphi^{(p)}(p^\infty; A) \to \overline{\mathbf{Q}}_p$ of $\operatorname{Spec}(h_\varphi^{(p)}(p^\infty; A))(\overline{\mathbf{Q}}_p)$ is called *algebraic* if it factors through $h_\varphi^{(p)}(p^r; A)$ for some finite r. Then the following theorem follows immediately from Propositions 1 and 2.

THEOREM 1 (WILES [Wi88]). *Let R be a local ring of $h_\varphi^{(p)}(p^\infty; A)$. If a λ-residual representation exists for every algebraic λ factoring through R, then there exist a pseudorepresentation $\pi'\colon G_F \to R$ and a continuous representation $\pi\colon G_F \to \operatorname{GL}_2(Q)$ such that $\operatorname{Tr}(\pi(\operatorname{Frob}_\mathscr{I})) = \operatorname{Tr}(\pi'(\operatorname{Frob}_\mathscr{I})) = T(\mathscr{I})$ and $\det(\pi(\operatorname{Frob}_\mathscr{I})) = \det(\pi'(\operatorname{Frob}_\mathscr{I})) = N(\mathscr{I})\langle\mathscr{I}\rangle$ for all prime ideals $\mathscr{I}$ prime to p, where Q is the total quotient ring of R.*

PROOF. Since G is unramified outside p, the set Σ of Frobenius elements for primes outside p is dense in G (Tchebotarev density theorem). We set $\operatorname{tr}(\operatorname{Frob}_q) = \lambda(T(q))$ and $\det(\operatorname{Frob}_q) = \chi(q)\kappa(\langle q \rangle)q^{-1}$. Let S be the subset of $\operatorname{Spec}(R)$ consisting of algebraic points. We identify $\lambda \in S$ with the prime ideal $P = \operatorname{Ker}(\lambda)$, i.e., $\lambda\colon R \to R/P$. We number each element of S and write $S = \{P_i\}_{i=1}^\infty$ and π_i for the residual representation attached to P_i. We construct out of each residual representation π_i for $P \in S$, a pseudorepresentation π_i'. Then all the π_i''s are compatible. Then by the above proposition, we can construct a pseudo-representation $\pi^{i'}$ into $R/\bigcap_{j=1}^{i} P_j$ so that

$$\mathrm{Tr}(\pi^{i'}(\sigma)) \equiv \mathrm{Tr}(\pi^{i-1'}(\sigma)) \bmod \bigcap_{j=1}^{i-1} P_j \quad \text{on } \Sigma.$$

Both sides of this congruence are continuous functions, and hence,

$$\mathrm{Tr}(\pi^{i'}(\sigma)) \equiv \mathrm{Tr}(\pi^{i-1'}(\sigma)) \bmod P_1 \cap \cdots \cap P_{i-1} \quad \text{on } G.$$

Note that, by definition, if $\pi'^i = (a_i, d_i, x_i)$, then $a_i(\sigma) = \frac{1}{2}(\mathrm{Tr}(\pi^{i'}(\sigma)) - \mathrm{Tr}(\pi^{i'}(\sigma c)))$, $d_i(\sigma) = \frac{1}{2}(\mathrm{Tr}(\pi^{i'}(\sigma)) + \mathrm{Tr}(\pi^{i'}(\sigma c)))$, and $x_i(\sigma, \tau) = a_i(\sigma\tau) - a_i(\sigma)a_i(\tau)$. Therefore, we have

$$a_i(\sigma) \equiv a_{i-1}(\sigma) \bmod P_1 \cap \cdots \cap P_{i-1}, \quad d_i(\sigma) \equiv d_{i-1}(\sigma) \bmod P_1 \cap \cdots \cap P_{i-1}$$

and

$$x_i(\sigma, \tau) \equiv x_{i-1}(\sigma, \tau) \bmod P_1 \cap \cdots \cap P_{i-1}.$$

Thus, we can define a pseudorepresentation π' into $R = \varprojlim_i R/P_1 \cap \cdots \cap P_i$ by

$$\pi'(\sigma) = \varprojlim_i \pi^{i'}(\sigma).$$

Then we can construct the representation π out of π' by Proposition 1, because we already know that Q is a direct sum of fields (i.e., R is reduced). $\square$

For any given continuous pseudorepresentation $\pi': G_F \to R$ for a profinite algebra R over $\mathbf{Z}_p$, there exists the largest closed normal subgroup $H(\pi')$ among closed normal subgroups H such that π' factors through G/H. In fact, $H(\pi')$ is the maximal closed normal subgroup in $\{\sigma \in G \mid \pi'(\sigma\eta) = \pi'(\sigma) = \pi'(\eta\sigma)$ for all $\sigma \in G\}$. If R is an integral domain and if there exists an absolutely irreducible Galois representation π attached to π', then $\mathrm{Ker}(\pi') = \mathrm{Ker}(\pi)$, because the isomorphism class of such representations over the quotient field of R is unique. Theorem 1 shows the existence of a big pseudorepresentation $\pi': G_F \to h_\varphi^{(p)}(p^\infty; A)$ such that $\mathrm{Tr}(\pi'(\mathrm{Frob}_\mathscr{I})) = T(\mathscr{I})$ and $\det(\pi'(\mathrm{Frob}_\mathscr{I})) = N(\mathscr{I})\langle\mathscr{I}\rangle$ for all prime ideals $\mathscr{I}$ prime to p, as long as there exist pseudorepresentations attached to algebraic points of $\mathrm{Spec}(h_\varphi^{(p)}(p^\infty; A))$. We write $F_{\mathrm{GL}(2),p}^{\mathrm{mod}}$ for the fixed field of $H(\pi')$ for the pseudorepresentation $\pi': G_F \to h_\varphi^{(p)}(p^\infty; A)$ if π' exists. This is one of the candidates for the maximal p-ramified GL(2)-extension. For each homomorphism $\lambda: h_\varphi^{(p)}(p^\infty; A) \to R$ of profinite A-algebras, we call a pseudorepresentation $\pi_\lambda: G_F \to R$ λ-residual if $\mathrm{Tr}(\pi_\lambda(\mathrm{Frob}_\mathscr{I})) = \lambda(T(\mathscr{I}))$ for all primes $\mathscr{I}$ prime to p.

To describe a more theoretical candidate for $F_{\mathrm{GL}(2),p}$ found by Mazur [**M89**], we assume the existence of a $\bar\lambda$-residual pseudorepresentation $\bar\pi$ for an algebra homomorphism $\bar\lambda: h_\varphi^{(p)}(p^\infty; A) \to \kappa$. In [**M89**], Mazur studied the universal deformation of each $\overline{\lambda}$-residual representation. Following his argument, we study here the universal deformation of each $\overline{\lambda}$-residual pseudorepresentation. We consider the functor

$$\mathscr{PR}_{\overline{\pi}}\colon \mathscr{C} \to \mathit{Sets} \quad \text{given by} \quad \mathscr{PR}_{\overline{\pi}}(R) = \{\pi \in \mathscr{PR}(R) \mid \pi \bmod m_R = \overline{\pi}\},$$

where m_R is the maximal ideal of R and $(\pi \bmod m_R) = (a', d', x')$ given by $a'(\sigma) = a(\sigma) \bmod m_R$, $d'(\sigma) = d(\sigma) \bmod m_R$, $x'(\sigma,\tau) = x(\sigma,\tau) \bmod m_R$ for $\pi = (a,d,x)$. Then we have

THEOREM 2. *The functor $\mathscr{PR}_{\overline{\pi}}$ is representable in $\mathscr{C}$. That is, there exists a unique pair $(R^{\mathrm{pr}}, \pi^{\mathrm{pr}})$ (up to isomorphisms) consisting of a Noetherian profinite A-algebra R^{pr} and a pseudorepresentation $\pi^{\mathrm{pr}}\colon G_F \to R^{\mathrm{pr}}$ such that*

$$\mathrm{Hom}_{\mathscr{C}}(R^{\mathrm{pr}}, R) \ni \varphi \mapsto \varphi \circ \pi^{\mathrm{pr}} \in \mathscr{PR}_{\overline{\pi}}(R)$$

induces an isomorphism $\mathrm{Hom}_{\mathscr{C}}(R^{\mathrm{pr}}, R) \cong \mathscr{PR}_{\overline{\pi}}(R)$ for all objects R of $\mathscr{C}$.

PROOF. Let $\mathscr{C}_0$ be the category of Artinian local A-algebras with residue field κ. We only need to show the prorepresentability of $\mathscr{PR}_{\overline{\pi}}$ restricted to $\mathscr{C}_0$. Let $\alpha_i\colon R_i \to R$ ($i = 1, 2$) be morphisms in $\mathscr{C}_0$, and let $\gamma\colon \mathscr{PR}_{\overline{\pi}}(R_1 \times_R R_2) \to \mathscr{PR}_{\overline{\pi}}(R_1) \times_{\mathscr{PR}_{\overline{\pi}}(R)} \mathscr{PR}_{\overline{\pi}}(R_2)$ be the natural map as in (11). To show the representability by a Noetherian ring in $\mathscr{C}$ of the covariant functor $\mathscr{PR}_{\overline{\pi}}$, we need to check the four criteria of Schlessinger [**Sch68**]. However, the three criteria denoted as (H_1), (H_2), and (H_4) in [**Sch68**] automatically follow from Proposition 2(i). The remaining criterion (H_3) follows immediately from

(F) $\qquad \mathscr{PR}_{\overline{\pi}}(\kappa[T]/(T^2))$ is a finite set.

We now prove (F). We write $\kappa[T]/(T^2)$ as $\kappa[\varepsilon]$ with $\varepsilon^2 = 0$ (i.e., ε is the image of T). Write $\overline{\pi} = (\overline{a}, \overline{d}, \overline{x})$, and take $\pi = (a, d, x) \in \mathscr{PR}_{\overline{\pi}}(\kappa[\varepsilon])$. If there exist $\rho, \eta \in G_F$ such that $\overline{x}(\rho, \eta) \neq 0$, then one defines a representation $\varphi\colon G_F \to \mathrm{GL}_2(\kappa)$ as in Proposition 1 using $\overline{x}(\rho, \eta)$. If there are no $\rho, \eta \in G_F$ as above, we set $\varphi(\sigma) = \begin{pmatrix} \overline{a}(\sigma) & 0 \\ 0 & \overline{d}(\sigma) \end{pmatrix}$. Set $H = \mathrm{Ker}(\varphi)$. If $\sigma \in H$, $\varphi(\sigma) = \begin{pmatrix} 1 & 0 \\ 0 & 1 \end{pmatrix}$, and therefore, $\overline{x}(\sigma, \eta) = \overline{x}(\rho, \sigma) = 0$ for all $\sigma \in H$. For all $\tau \in G$, we have by (PR4) that $\overline{x}(\sigma, \tau)\overline{x}(\rho, \eta) = \overline{x}(\sigma, \eta)\overline{x}(\rho, \tau) = 0$. This shows $\overline{x}(\sigma, \tau) = 0$ for all $\sigma \in H$ because $\overline{x}(\rho, \eta) \neq 0$. From $\overline{x}(\tau, \sigma)\overline{x}(\rho, \eta) = \overline{x}(\tau, \eta)\overline{x}(\rho, \sigma) = 0$, we conclude

(12a) $\qquad \overline{x}(\sigma, \tau) = \overline{x}(\tau, \sigma) = 0 \quad \text{for all } \tau \in G \text{ and } \sigma \in H.$

We also have

(12b) $\qquad \overline{a}(\sigma) = \overline{d}(\sigma) = 1 \quad \text{if } \sigma \in H.$

Then by (PR2), we have

(12c) $\qquad \begin{aligned} x(\sigma\tau, \delta\gamma) &= a(\sigma)a(\gamma)x(\tau,\delta) + a(\gamma)d(\tau)x(\sigma,\delta) \\ &\quad + a(\sigma)d(\delta)x(\tau,\gamma) + d(\tau)d(\delta)x(\sigma,\gamma). \end{aligned}$

We see, observing $x(1, *) = x(*, 1) = 0$, that

$$x(\sigma, \delta\gamma) = a(\gamma)x(\sigma,\delta) + d(\delta)x(\sigma,\gamma) \qquad (\tau = 1 \text{ in (12c)}),$$
$$x(\sigma\tau, \delta) = a(\sigma)x(\tau,\delta) + d(\tau)x(\sigma,\delta) \qquad (\gamma = 1 \text{ in (12c)}).$$

Therefore by (12a, b, c), we have, writing $a = \overline{a} \oplus a'\varepsilon$, $d = \overline{d} \oplus d'\varepsilon$, and $x = \overline{x} \oplus x'\varepsilon$,

$$x'(\sigma, \delta\gamma) = x'(\sigma, \delta) + x'(\sigma, \gamma) \quad \text{if } \delta, \gamma \in H \text{ and } \sigma \in G,$$
$$x'(\sigma\tau, \delta) = x'(\sigma, \delta) + x'(\tau, \delta) \quad \text{if } \sigma, \tau \in H \text{ and } \delta \in G.$$

Therefore, for each $\sigma \in G$ the maps $_\sigma x\colon H \ni \delta \mapsto x'(\sigma, \delta) \in \kappa$ and $x_\sigma\colon H \ni \delta \mapsto x'(\delta, \sigma) \in \kappa$ are homomorphisms of groups. Thus, if we write H' for the closed subgroup of H topologically generated by $H^p(H, H)$, then $\#(H/H') < +\infty$ by class field theory, and $x'(\sigma, \tau) = 0$ for all $\sigma, \tau \in H'$. By (12a),

(12d) $\quad x(\sigma, \tau) = 0 \quad \text{and} \quad x(\delta\sigma, \gamma\tau) = a(\tau)d(\sigma)x(\delta, \gamma) \quad \text{for all } \sigma, \tau \in H'.$

Then by (12d) and (PR2), we see that

(12e) $\quad a'(\sigma\tau) = a'(\sigma) + a'(\tau), \qquad d'(\sigma\tau) = d'(\sigma) + d'(\tau) \quad \text{for all } \sigma, \tau \in H'.$

Let H'' be the subgroup of H' topologically generated by $H'^p(H', H')$. Then again, we have $\#(H/H'') < +\infty$ and

(12f) $\quad\quad\quad\quad\quad\quad a'(H'') = 0 \quad \text{and} \quad d'(H'') = 0.$

Then we have

(12g) $\quad a(\gamma\sigma) = a(\gamma), \qquad d(\gamma\sigma) = d(\gamma), \quad \text{and} \quad x(\delta\sigma, \gamma\tau) = x(\delta, \gamma) \quad \text{if } \sigma, \tau \in H''.$

Note that H'' does not depend on the choice of π. Therefore, all the elements of $\mathscr{PR}_{\overline{\pi}}(\kappa[\varepsilon])$ are functions of the finite group G/H'', and thus, $\#(\mathscr{PR}_{\overline{\pi}}(\kappa[\varepsilon])) < +\infty$. $\square$

PROPOSITION 3. R^{pr} is generated by $\text{Tr}(\pi^{\text{pr}})(G_F)$ over A.

PROOF. Let R_{tr} be the A-subalgebra topologically generated by $\text{Tr}(\pi^{\text{pr}}(G_F))$. Since π^{pr} is determined by its trace, π^{pr} has values in R_{tr}. Therefore, we have a deformation $(R_{\text{tr}}, \pi^{\text{pr}})$. For each deformation (A, π) of $(\kappa, \overline{\pi})$, we have a morphism $\varphi_\pi\colon R_{\text{tr}} \to A$ such that $\pi = \varphi_\pi \circ \pi^{\text{pr}}$. The morphism φ_π is unique because R_{tr} is generated by the traces. Therefore, $(R_{\text{tr}}, \pi^{\text{pr}})$ is already universal and hence $R^{\text{pr}} = R_{\text{tr}}$. $\square$

Let $\overline{\rho}$ be a $\overline{\lambda}$-residual representation. In [M89], Mazur considered instead of $\mathscr{PR}_{\overline{\pi}}$ the following functor

$$\mathscr{R}_{\overline{\rho}}(R) = \{\rho\colon G \to \text{GL}_2(R) \mid \rho \bmod \mathfrak{m}_R = \overline{\rho}\}/\approx,$$

where ρ is assumed to be a continuous homomorphism and $\rho \approx \rho'$ if there exists a matrix $\alpha \in \text{GL}_2(R)$ with $\alpha \equiv \begin{pmatrix} 1 & 0 \\ 0 & 1 \end{pmatrix} \bmod \mathfrak{m}_R$ such that $\rho(\sigma) = \alpha\rho'(\sigma)\alpha^{-1}$ for all $\sigma \in G$. Then Mazur proved

THEOREM 3 (Mazur). If $\overline{\rho}$ is absolutely irreducible, then $\mathscr{R}_{\overline{\rho}}$ is representable by a unique pair $(R^{\text{rep}}, \rho^{\text{rep}})$ up to isomorphisms for a Noetherian local ring R^{rep} in $\mathscr{C}$ and $\rho^{\text{rep}} \in \mathscr{R}_{\overline{\rho}}(R^{\text{rep}})$.

It is easy to show

COROLLARY 2. Let $\overline{\pi}$ be the pseudorepresentation attached to $\overline{\rho}$. Then R^{pr} is canonically isomorphic to the subalgebra of R^{rep} topologically generated by traces of ρ^{rep} over A.

Assuming that there are only finitely many connected components of $h_\varphi^{(p)}(p^\infty; A)$ and that Questions 1 and 2 have affirmative answers, we write $\mathbf{h}$ (resp., $\mathbf{h}^{(p)}$) for

$h_\varphi(p^\infty; A)$ (resp., $h_\varphi^{(p)}(p^\infty; A)$). We further assume, extending by scalar if necessary, that all the local rings of $\mathbf{h}^{(p)}$ have κ as their residue field. We further assume that there exists a pseudorepresentation $\pi\colon G_F \to \mathbf{h}$ such that $\mathrm{Tr}(\pi)(\mathrm{Frob}_\mathscr{J}) = T(\mathscr{J})$ for all primes $\mathscr{J}$ prime to p. Let π_0 be the set of local rings of $\mathbf{h}^{(p)}$ ($=$ the set of connected components of $\mathrm{Spec}(\mathbf{h}^{(p)})$). For each $S \in \pi_0$, we have a pseudorepresentation $\bar\pi_S = \pi \bmod m_S$. To $\bar\pi_S$ we can attach, by Theorem 2, the universal deformation $(R_S^{\mathrm{pr}}, \pi_S^{\mathrm{pr}})$. Let π_S be the projected image of π to S. Then (S, π_S) is a deformation of $\bar\pi_S$, and hence, there exists a natural ring homomorphism $\varphi_S\colon R_S^{\mathrm{pr}} \to S$ satisfying $\varphi_S \circ \pi_S^{\mathrm{pr}} = \pi_S$. We set $\mathbf{H} = \bigoplus_{S \in \pi_0} R_S^{\mathrm{pr}}$. Then we have a surjective ring homomorphism $\Phi\colon \mathbf{H} \to \mathbf{h}^{(p)}$. Then by Proposition 3, the following fact is clear:

COROLLARY 3. *The ring homomorphism* $\Phi\colon \mathbf{H} \to \mathbf{h}^{(p)}$ *is surjective, and hence,* $\mathbf{h}^{(p)}$ *is Noetherian. The algebra* $\mathbf{h}$ *is generated topologically over* $\mathbf{h}^{(p)}$ *by* $T(\varpi_\mathfrak{p})$ *for all prime ideals* $\mathfrak{p}$ *above* p *in* F, *and hence, is Noetherian, where* $\varpi_\mathfrak{p}$ *is a prime element in* $F_\mathfrak{p}$.

Then we ask in the spirit of [**M89**] and [**MT90**]

QUESTION 5. *Is* Φ *injective?*

We can think of the field $F_{\mathrm{GL}(2),p}$ fixed by $\mathrm{Ker}(\bigoplus_S \pi_S^{\mathrm{pr}})$, which is an extension of $F_{\mathrm{GL}(2),p}^{\mathrm{mod}}$. Then Question 5 is just asking whether or not $F_{\mathrm{GL}(2),p}^{\mathrm{mod}} = F_{\mathrm{GL}(2),p}$.

§8

In the discussion of the previous paragraph, we only used the Hecke operators $T(z)$ with $z_p = 1$. The existence of $T(p)$ in $\mathbf{h}_\varphi(p^\infty; A)$ plays a key role in the definition of the ordinary part of $\mathbf{h}_\varphi(p^\infty; A)$, which is the most manageable part of the algebra. Since $\mathbf{h}_\varphi(p^\infty; A)$ is a profinite algebra, we can find two idempotents e_h and $e^h = 1 - e_h$ in $\mathbf{h}_\varphi(p^\infty; A)$ for any given element h in $\mathbf{h}_\varphi(p^\infty; A)$ such that $e_h h \in (e_h \mathbf{h}_\varphi(p^\infty; A))^\times$ and $\lim_{n \to \infty}(e^h h)^n = 0$ (see [**H89a**], §4] and [**H93b**], §7.2]). We apply this argument to $h = T(p)$. We write $e = e_{T(p)}$ and put $\mathbf{h}_\varphi^{n.\,\mathrm{ord}} = \mathbf{h}_\varphi^{n.\,\mathrm{ord}}(p^\infty; A) = e\mathbf{h}_\varphi(p^\infty; A)$. Similarly, we define $h_\varphi^{n.\,\mathrm{ord}} = h_\varphi^{n.\,\mathrm{ord}}(p^\infty; A)$ out of $h_\varphi(p^\infty; A)$ and $T(p)$. We write again $T(z)$ and $\langle z \rangle \in \mathbf{h}_\varphi^{n.\,\mathrm{ord}}$ for $eT(z)$ and $e\langle z \rangle$. Then we can prove [**H93a**], [**H94b**]:

THEOREM 4. *The triple* $\{\mathbf{h}_\varphi^{n.\,\mathrm{ord}}(p^\infty; A)_{/F}, (T(z), \langle z \rangle)_{z \in \mathscr{Z}}\}$ *is independent of* φ (*as long as* $n(\varphi) \neq 0$). *The same assertion holds for* $\{h_\varphi^{n.\,\mathrm{ord}}(p^\infty; A)_{/F}, (T(z), \langle z \rangle)\}$ *if* F *is totally real.*

We write $(\mathbf{h}_{/F}^{n.\,\mathrm{ord}}, T(z), \langle z \rangle)$ (resp., $(h_{/F}^{n.\,\mathrm{ord}}, T(z), \langle z \rangle)$) for the universal triple $(\mathbf{h}_\varphi^{n.\,\mathrm{ord}}(p^\infty; A)_{/F}, T(z), \langle z \rangle)$ (resp., $(h_\varphi^{n.\,\mathrm{ord}}(p^\infty; A)_{/F}, T(z), \langle z \rangle)$ when F is totally real). A point $P \in \mathrm{Hom}_{A\text{-alg}}(\mathbf{h}^{n.\,\mathrm{ord}}, \overline{\mathbf{Q}}_p) = \mathrm{Spec}(h^{n.\,\mathrm{ord}})(\overline{\mathbf{Q}}_p)$ is called *arithmetic* if the composition $\varphi(P) = P \circ (T \times \langle\ \rangle)\colon \mathbf{G} \to \overline{\mathbf{Q}}_p$ is arithmetic. Of course, $\varphi(P)$ may be different from φ. We write $r(P) = r(\varphi(P))$, $n(P) = n(\varphi(P))$, and $v(P) = v(\varphi(P))$. Then

THEOREM 5. *Each arithmetic point* $P \in \mathrm{Spec}(\mathbf{h}^{n.\,\mathrm{ord}})(\overline{\mathbf{Q}}_p)$ *factors through* $h_{\varphi(P)}^{n.\,\mathrm{ord}}(p^\infty; A)$. *For each arithmetic point* $P \in \mathrm{Hom}_{A\text{-alg}}(\mathbf{h}^{n.\,\mathrm{ord}}, \overline{\mathbf{Q}}_p)$, *we have* $P(T(\mathscr{J})) \in \overline{\mathbf{Q}}$ *and* $P(\langle \mathscr{J} \rangle) \in \overline{\mathbf{Q}}$ *for all prime ideals* $\mathscr{J}$ *prime to* p, *and there exist a Hecke* *character ξ whose infinity type is given by $n(P)+2v(P)$ and a unique algebraic cuspidal automorphic representation $\pi(P)$ occurring in $L_2^0(\xi)$ such that $\pi(P) = \otimes_{\mathscr{I}} \pi_{\mathscr{I}}(P)$ and*

$$L(s,\pi_{\mathscr{I}}(P))^{-1} = 1 - P(T(\mathscr{I}))N(\mathscr{I})^{-s} + P(\langle\mathscr{I}\rangle)N(\mathscr{I})^{1-2s} \text{ for all } \mathscr{I} \text{ prime to } p.$$

When F is totally real, the above two theorems are proven in [**H89b**, 2.5] (see also [**H89a**, §4] and [**H93b**, §7.3]). In general, the results follow from the main result in [**H93a**] (see [**H94b**] for the proof).

Let $L(s,\pi(P))$ be the standard L-function of $\pi(P)$ introduced in §3. Let $\pi(P)^\vee$ be the contragredient representation of $\pi(P)$. For two arithmetic points P and Q, we consider the external tensor product representation $\pi(P) \times \pi(Q)^\vee$ of $\mathrm{GL}_2(F_A) \times \mathrm{GL}_2(F_A)$. We write $L(s,\pi(P) \times \pi(Q)^\vee)$ for the standard L-function of $\pi(R) \times \pi(Q)^\vee$ [**H91a**]. Writing

$$L(s,\pi_{\mathscr{I}}(P)) = \{(1-\alpha N(\mathscr{I})^{-s})(1-\beta N(\mathscr{I})^{-s})\}^{-1}$$

and

$$L(s,\pi_{\mathscr{I}}(Q)^\vee) = \{(1-\alpha' N(\mathscr{I})^{-s})(1-\beta' N(\mathscr{I})^{-s})\}^{-1},$$

we have the following local Euler factor:

$$L(s,\pi_{\mathscr{I}}(P) \times \pi_{\mathscr{I}}(Q)^\vee)$$
$$= \{(1-\alpha\alpha' N(\mathscr{I})^{-s})(1-\alpha\beta' N(\mathscr{I})^{-s})(1-\beta\alpha' N(\mathscr{I})^{-s})(1-\beta\beta' N(\mathscr{I})^{-s})\}^{-1}.$$

Write $L(s)$ for one of the above L-functions and $L_\infty(s)$ for the Γ-factor of $L(s)$. We suppose that the functional equation of $L(s)$ is given by $s \leftrightarrow w+1-s$. Then w is an integer. We call $L(1)$ critical if $L_\infty(s)$ is finite at $s=1$ and $s=w$. We call $\pi(P)$ (resp., $\pi(P) \times \pi(Q)^\vee$) critical, if $L(1,\pi(P))$ (resp., $L(1,\pi(P) \times \pi(Q)^\vee)$) is critical. Note that our automorphic representations $\pi(P)$ and $\pi(Q)$ are algebraic. Thus, there should exist corresponding motives $M(P)$ and $M(Q)$ (see [**BR93**]). If $\pi(P)$ (resp., $\pi(P) \times \pi(Q)^\vee$) is critical, then $M(P)(1)$ (resp., $M(P) \otimes M(Q)^\vee(1)$) is critical in the sense of Deligne [**D79**], where $M(Q)^\vee$ is the dual of $M(Q)$. Thus, we have well-defined motivic periods $c^+(M(P)(1))$ and $c^+(M(P) \otimes M(Q)^\vee(1))$, which are nonzero complex numbers. We write $c^+(P)$ (resp., $c^+(P,Q)$) for the identity component of $c^+(M(P)(1))$ (resp., $c^+(M(P) \otimes M(Q)^\vee(1))$. Then the algebraicity conjecture [**D**] tells us

$$\frac{L(1,\pi(P))}{c^+(P)} \in \mathbf{Q}(\pi(P)) \quad \text{and} \quad \frac{L(1,\pi(P) \times \pi(Q)^\vee)}{c^+(P,Q)} \in \mathbf{Q}(\pi(P) \otimes \pi(Q)^\vee),$$

where $\mathbf{Q}(\pi(P))$ (resp., $\mathbf{Q}(\pi(P) \otimes \pi(Q)^\vee)$) is the number field generated by $P(T(\mathscr{I}))$ (resp., $P(T(\mathscr{I}))$ and $Q(T(\mathscr{I}))$) for all $\mathscr{I}$ prime to p (which is the field of definition of the finite part of $\pi(P)$ (resp., $\pi(P) \otimes \pi(Q)^\vee$).

If an element L of the total quotient ring of $\mathbf{h}^{n.\,\mathrm{ord}}$ (resp., $\mathbf{h}^{n.\,\mathrm{ord}} \widehat{\otimes}_A \mathbf{h}^{n.\,\mathrm{ord}}$) is given, then we can think of L as a (p-adic meromorphic) function on $\mathrm{Spec}(\mathbf{h}^{n.\,\mathrm{ord}})(\overline{\mathbf{Q}}_p)$ (resp., $\mathrm{Spec}(\mathbf{h}^{n.\,\mathrm{ord}}) \times \mathrm{Spec}(\mathbf{h}^{n.\,\mathrm{ord}})(\overline{\mathbf{Q}}_p)$) by $L(P) = P(L)$ (resp., $L(P,Q) = P\widehat{\otimes}Q(L)$) as long as the value is well defined. Here we have regarded $P\colon \mathbf{h}^{n.\,\mathrm{ord}} \to \overline{\mathbf{Q}}_p$ (resp., $P\widehat{\otimes}Q\colon \mathbf{h}^{n.\,\mathrm{ord}} \widehat{\otimes}_A \mathbf{h}^{n.\,\mathrm{ord}} \to \overline{\mathbf{Q}}_p$) as an A-algebra homomorphism, and if sL for $s \in \mathbf{h}^{n.\,\mathrm{ord}}$ (resp., $s \in \mathbf{h}^{n.\,\mathrm{ord}} \widehat{\otimes}_A \mathbf{h}^{n.\,\mathrm{ord}}$) is in $\mathbf{h}^{n.\,\mathrm{ord}}$ (resp., $\mathbf{h}^{n.\,\mathrm{ord}} \widehat{\otimes}_A \mathbf{h}^{n.\,\mathrm{ord}}$) and $P(s) \neq 0$ (resp.,

$P\widehat{\otimes}Q(s) \neq 0$), then $P(L) = P(sL)/P(s)$ (resp., $P\widehat{\otimes}Q(L) = (P\widehat{\otimes}Q(sL))/(P\widehat{\otimes}Q(s))$). Then we can ask

QUESTION 6. Can we find a p-adic meromorphic function $L_p(P)$ (resp., $L_p(P,Q)$) on $\mathrm{Spec}(\mathbf{h}^{n.\,\mathrm{ord}})(\overline{\mathbf{Q}}_p)$ (resp., $\mathrm{Spec}(\mathbf{h}^{n.\,\mathrm{ord}}\widehat{\otimes}_A\mathbf{h}^{n.\,\mathrm{ord}})(\overline{\mathbf{Q}}_p)$) interpolating the values $L(1,\pi(P))/c^+(P) \in \mathbf{Q}(\pi(P))$ (resp., $L(1,\pi(P)\times\pi(Q)^\vee)/c^+(P,Q) \in \mathbf{Q}(\pi(P)\otimes\pi(Q)^\vee)$) as long as $\pi(P)$ (resp., $\pi(P)\times\pi(Q)^\vee$) is critical.

When F is totally real, the algebraicity conjecture is known for $L(1,\pi(P))$ by Mazur and Manin (see [H93b], §6.4]) and also by Shimura [Sh88, 90] and for $L(1,\pi(P)\times\pi(Q)^\vee)$ by Shimura [Sh88] (see also [H93b, §10.4]). When F has complex places, an approximation of the algebraicity conjecture for $L(s,\pi(P))$ is known ([H94a]). When F contains CM-fields, a partial result is given for the algebraicity conjecture for $L(1,\pi(P)\times\pi(Q)^\vee)$ in [H94a]. When $F \neq \mathbf{Q}$, the transcendental factors vary depending on $n(P)$ and $v(P)$ for $L(1,\pi(P))$ and on $n(P), n(Q), v(P)$, and $v(Q)$ for $L(1,\pi(P)\times\pi(Q)^\vee)$ (cf. [Sh88], [H94a]). Thus, when $F \neq \mathbf{Q}$, there might be several p-adic L-functions interpolating different combinations of special values. In any case, the answer to the above question is affirmative when $F = \mathbf{Q}$ ([Ki91], [GS92], and [H93b], §§7.4 and 10.4]). When F is totally real, the answer is known, so far, to be partially affirmative [H91]. We can ask similar questions for various Langlands L-functions of the polynomial representations of the L-group of $\mathrm{GL}(2)_{/F}$. For example, the adjoint lift p-adic L-function exists on $\mathrm{Spec}(\mathbf{h}^{n.\,\mathrm{ord}}_{/\mathbf{Q}})$ [H90a].

As for Question 4, we have the following partial answer.

THEOREM 6. *The algebra $\mathbf{h}^{n.\,\mathrm{ord}}$ is of finite type over $A[[W]]$. If F is totally real and $h^{n.\,\mathrm{ord}} \neq 0$, then $h^{n.\,\mathrm{ord}}$ is a torsion-free $A[[W]]$-module. In particular, in this case the natural map $\mathrm{Spec}(h^{n.\,\mathrm{ord}}) \to \mathrm{Spec}(A[[W]])$ is dominant. If F has some complex places, then the image of $\mathrm{Spec}(h^{n.\,\mathrm{ord}})$ is contained in a proper closed subscheme of $\mathrm{Spec}(A[[W]])$.*

The above fact is proven in [H89b] for totally real fields F. Thus, the p-adic space attached to each irreducible component of $h^{n.\,\mathrm{ord}}$ has dimension equal to $[F:\mathbf{Q}]+1+\delta(p,F)$, which grows at least linearly with respect to the degree $[F:\mathbf{Q}]$. For general number fields, it follows from the main result of [H93a] (see [H94b] for the proof). Thus, we may ask

QUESTION 7. Characterize the image of $\mathrm{Spec}(h^{n.\,\mathrm{ord}})$ in $\mathrm{Spec}(A[[W]])$.

If F is totally real and if the Leopoldt conjecture is true for F and p, the dimension of the scheme $\mathrm{Spec}(h^{n.\,\mathrm{ord}}) \times_{\mathbf{Z}_p} \mathbf{Q}_p$ is $[F:\mathbf{Q}]+1$. In particular, it is equal to 2 when $F = \mathbf{Q}$. This value coincides with the dimension of the continuous spectrum in the complex case. We may thus ask

QUESTION 8. Is the dimension (over $\mathbf{Q}_p$) of the irreducible components of the p-adic nearly ordinary Hecke algebra for $\mathrm{GL}(n)_{/\mathbf{Q}}$ (suitably defined) equal to the maximum of the dimension of the continuous complex spectrum for $\mathrm{GL}(n)_{/\mathbf{Q}}$ (which is equal to n)?

It is quite plausible that the dimension is less than or equal to n. However, there is no compelling reason to expect that the answer is exactly n. The final point we would like to make explicit is

QUESTION 9. What kind of arithmetic of $F_{\mathrm{GL}(2),p}^{\mathrm{mod}}$ do the above p-adic L-functions describe?

For the connected component of $\mathrm{Spec}(h^{n.\,\mathrm{ord}})$ corresponding to the representation of Weil type (§6), there might be some chance to relate these p-adic L-functions to a certain main conjecture of the Iwasawa theory. Partial results are obtained in this direction when F is totally real ([**MT90**], [**Ti89**], [**HT91**], [**HT93a, b**]). When the representation attached to the connected component is not of Weil type, there are some speculations but their meaning is not yet clear [**MT90**].

References

[BR89]   D. Blasius and J. D. Rogawski, *Galois representations for Hilbert modular forms*, Bull. Amer. Math. Soc. **21** (1989), 65–69.

[BR93]   _____, *Motives for Hilbert modular forms*, Invent. Math. **114** (1993), 55–87.

[C90]   L. Clozel, *Motifs et formes automorphes: Applications du principe de fonctorialité*, Automorphic Forms, Shimura Varieties, and L-functions, Perspectives in Math., vol. 10, Academic Press, New York, 1990, pp. 77–159.

[D79]   P. Deligne, *Valeurs de fonctions L et périodes d'intégrales*, Proc. Sympos. Pure Math., vol. 33, part 2, Amer. Math. Soc., Providence, R.I., 1979, pp. 313–346.

[DMOS82]   P. Deligne, J. S. Milne, A. Ogus, and K-y Shih, *Hodge cycles, motives and Shimura varieties*, Lecture Notes in Math., vol. 900, Springer-Verlag, Berlin and New York, 1982.

[GJ72]   R. Godement and H. Jacquet, *Zeta functions of simple algebras*, Lecture Notes in Math., vol. 260, Springer-Verlag, Berlin and New York, 1972.

[Gv88]   F. Q. Gouvêa, *Arithmetic of p-adic modular forms*, Lecture Notes in Math., vol. 1304, Springer-Verlag, Berlin and New York, 1988.

[Gv90]   _____, *Deforming Galois representations: Controlling the conductor*, J. Number Theory **34** (1990), 95–113.

[GS92]   R. Greenberg and G. Stevens, *p-adic L-functions and p-adic periods of modular forms*, preprint.

[Ha87]   G. Harder, *Eisenstein cohomology of arithmetic groups, the case* GL_2, Invent. Math. **89** (1987), 37–118.

[He85]   G. Henniart, *Le point sur la conjecture de Langlands pour* $\mathrm{GL}(n)$, Sém. Théorie des Nombres, Paris **1983–84** (1985), 115–131.

[H89a]   _____, *Theory of p-adic Hecke algebras and Galois representations*, Sūgaku **39** (1987), 124–139; English transl., Sūgaku Expositions **2** (1989), 75–102.

[H89b]   _____, *On nearly ordinary Hecke algebras for* $\mathrm{GL}(2)$ *over totally real fields*, Adv. Stud. Pure Math., vol. 17, Academic Press, Boston, MA, 1989, pp. 139–169.

[H89c]   _____, *Nearly ordinary Hecke algebras and Galois representations of several variables*, in supplement to the Amer. J. Math., Algebraic Analysis, Geometry and Number Theory, Johns Hopkins Univ. Press, Baltimore, MD, 1989, pp. 115–134.

[H90a]   _____, *p-adic L-functions for base change lifts of* GL_2 *to* GL_3, Proceedings of the Conference on Automorphic Forms, Shimura Varieties, and L-functions, Ann Arbor, Michigan, July 1988, Perspectives in Math., vol. 11, Academic Press, Boston, MA, 1990, pp. 93–142.

[H90b]   _____, *Le produit de Petersson et de Rankin p-adique*, Sém. Théorie de Nombre de Paris **1988–89** (1990), 87–102.

[H91]   _____, *On p-adic L-functions of* $\mathrm{GL}(2) \times \mathrm{GL}(2)$ *over totally real fields*, Ann. Institut Fourier **41**, No. 2 (1991), 311–391.

[H93a]   _____, *p-ordinary cohomology groups for* $\mathrm{SL}(2)$ *over number fields*, Duke Math. J. **69** (1993), 259–314.

[H93b]   _____, *Elementary theory of L-functions and Eisenstein series*, Student Texts in Mathematics, No. 26, Cambridge University Press, London and New York, 1993.

[H94a]   H. Hida, *On the critical values of L-functions of* $\mathrm{GL}(2)$ *and* $\mathrm{GL}(2) \times \mathrm{GL}(2)$, Duke Math. J. **74** (1994), 431–530.

[H94b]   _____, *p-adic ordinary Hecke algebras for* $\mathrm{GL}(2)$, Ann. Inst. Fourier (to appear).

[HT91] H. Hida and J. Tilouine, *Katz p-adic L-functions, congruence modules and deformation of Galois representations*, Proceedings of the London Math. Soc. Symposium on L-functions and arithmetic, Durham, England, July 1989, London Math. Soc. Lecture Note Ser., vol. 153, Cambridge Univ. Press, London and New York, 1991, pp. 271–293.

[HT93a] _____, *Anti-cyclotomic Katz p-adic L-functions and congruence modules*, Ann. Sci. École Norm. Sup. (4) **26** (1993), 189–259.

[HT93b] _____, *On the anticyclotomic main conjecture for CM fields*, Invent. Math. **117** (1994), 89–147.

[I72] K. Iwasawa, *Lectures on p-adic L-functions*, Ann. of Math. Stud., vol. 74, Princeton University Press, Princeton, NJ, 1972.

[K75] N. M. Katz, *p-adic L-functions via moduli of elliptic curves*, Proc. Sympos. Pure Math., vol. 29, Amer. Math. Soc., Providence, RI, 1975, pp. 479–506.

[K78] _____, *p-adic L-functions for CM fields*, Invent. Math. **49** (1978), 199–297.

[Ki91] K. Kitagawa, *On standard p-adic L-functions of families of elliptic cusp forms*, Ph.D. thesis, Univ. of California, Los Angeles, CA, 1991 (to appear in Contemporary Math.).

[Ku87] P. C. Kutzko, *On the supercuspidal representation of GL_N and other p-adic groups*, Proc. Internat. Congress Math. 1986, Berkeley, I, Amer. Math. Soc., Providence, RI, 1987, pp. 853–861.

[L76] R. P. Langlands, *On the functional equations satisfied by Eisenstein series*, Lecture Notes in Math., vol. 544, Springer-Verlag, Berlin and New York, 1976.

[M89] B. Mazur, *Deforming Galois representations*, Galois Groups over $\mathbf{Q}$, Springer-Verlag, Berlin and New York, 1989, pp. 385–437.

[MT90] B. Mazur and J. Tilouine, *Représentations Galoisiennes, différentielles de Kähler et conjectures principales*, Inst. Hautes Études Sci. Publ. Math. **71** (1990), 65–103.

[Mi89] T. Miyake, *Modular forms*, Springer-Verlag, Berlin and New York, 1989.

[N86] J. Neukirch, *Class field theory*, Grundlehren Math. Wiss. 280, Springer, 1986.

[Sch68] M. Schlessinger, *Functors on Artin rings*, Trans. Amer. Math. Soc. **130** (1968), 208–222.

[Se75] J.-P. Serre, *Valeurs propres des operateurs de Hecke modulo ℓ*, Astérisque **24–25** (1975), 109–117.

[Sh71] G. Shimura, *Introduction to the arithmetic theory of automorphic functions*, Iwanami Shoten and Princeton University Press, Princeton, NJ, 1971.

[Sh88] _____, *On the critical values of certain Dirichlet series and the periods of automorphic forms*, Invent. Math. **94** (1988), 245–305.

[Sh90] G. Shimura, *On the fundamental periods of automorphic forms of arithmetic type*, Invent. Math. **102** (1990), 399–428.

[Ta89] R. Taylor, *On Galois representations associated to Hilbert modular forms*, Invent. Math. **98** (1989), 265–280.

[Ta] _____, *Congruences between modular forms on GL_2 over imaginary quadratic fields*, preprint.

[T79] J. Tate, *Number theoretic background*, Proc. Sympos. Pure Math., vol. 33, part 2, Amer. Math. Soc., Providence, RI, 1979, pp. 3–26.

[Ti89] J. Tilouine, *Sur la conjecture principale anticyclotomique*, Duke Math. J. **59** (1989), 629–673.

[Wa82] L. C. Washington, *Introduction to cyclotomic fields*, Graduate Texts in Math., vol. 83, Springer, New York, 1982.

[W84] A. Weil, *Number Theory, An approach through history from Hammurapi to Legendre*, Birkhäuser, Boston, 1984.

[W51] _____, *Sur la théorie du corps de classes*, J. Math. Soc. Japan **3** (1951), 1–35.

[Wi88] A. Wiles, *On ordinary λ-adic representations associated to modular forms*, Invent. Math. **94** (1988), 529–573.

DEPARTMENT OF MATHEMATICS, UNIVERSITY OF CALIFORNIA AT LOS ANGELES, LOS ANGELES, CA 90024

Translated by HARUZO HIDA

Recent Titles in This Series

(Continued from the front of this publication)

121 V. D. Mazurov, Yu. I. Merzlyakov, and V. A. Churkin, Editors, The Kourovka Notebook: Unsolved Problems in Group Theory
120 M. G. Kreĭn and V. A. Jakubovič, Four Papers on Ordinary Differential Equations
119 V. A. Dem′janenko et al., Twelve Papers in Algebra
118 Ju. V. Egorov et al., Sixteen Papers on Differential Equations
117 S. V. Bočkarev et al., Eight Lectures Delivered at the International Congress of Mathematicians in Helsinki, 1978
116 A. G. Kušnirenko, A. B. Katok, and V. M. Alekseev, Three Papers on Dynamical Systems
115 I. S. Belov et al., Twelve Papers in Analysis
114 M. Š. Birman and M. Z. Solomjak, Quantitative Analysis in Sobolev Imbedding Theorems and Applications to Spectral Theory
113 A. F. Lavrik et al., Twelve Papers in Logic and Algebra
112 D. A. Gudkov and G. A. Utkin, Nine Papers on Hilbert's 16th Problem
111 V. M. Adamjan et al., Nine Papers on Analysis
110 M. S. Budjanu et al., Nine Papers on Analysis
109 D. V. Anosov et al., Twenty Lectures Delivered at the International Congress of Mathematicians in Vancouver, 1974
108 Ja. L. Geronimus and Gábor Szegő, Two Papers on Special Functions
107 A. P. Mišina and L. A. Skornjakov, Abelian Groups and Modules
106 M. Ja. Antonovskiĭ, V. G. Boltjanskiĭ, and T. A. Sarymsakov, Topological Semifields and Their Applications to General Topology
105 R. A. Aleksandrjan et al., Partial Differential Equations, Proceedings of a Symposium Dedicated to Academician S. L. Sobolev
104 L. V. Ahlfors et al., Some Problems on Mathematics and Mechanics, On the Occasion of the Seventieth Birthday of Academician M. A. Lavrent′ev
103 M. S. Brodskiĭ et al., Nine Papers in Analysis
102 M. S. Budjanu et al., Ten Papers in Analysis
101 B. M. Levitan, V. A. Marčenko, and B. L. Roždestvenskiĭ, Six Papers in Analysis
100 G. S. Ceĭtin et al., Fourteen Papers on Logic, Geometry, Topology and Algebra
99 G. S. Ceĭtin et al., Five Papers on Logic and Foundations
98 G. S. Ceĭtin et al., Five Papers on Logic and Foundations
97 B. M. Budak et al., Eleven Papers on Logic, Algebra, Analysis and Topology
96 N. D. Filippov et al., Ten Papers on Algebra and Functional Analysis
95 V. M. Adamjan et al., Eleven Papers in Analysis
94 V. A. Baranskiĭ et al., Sixteen Papers on Logic and Algebra
93 Ju. M. Berezanskiĭ et al., Nine Papers on Functional Analysis
92 A. M. Ančikov et al., Seventeen Papers on Topology and Differential Geometry
91 L. I. Barklon et al., Eighteen Papers on Analysis and Quantum Mechanics
90 Z. S. Agranovič et al., Thirteen Papers on Functional Analysis
89 V. M. Alekseev et al., Thirteen Papers on Differential Equations
88 I. I. Eremin et al., Twelve Papers on Real and Complex Function Theory
87 M. A. Aĭzerman et al., Sixteen Papers on Differential and Difference Equations, Functional Analysis, Games and Control
86 N. I. Ahiezer et al., Fifteen Papers on Real and Complex Functions, Series, Differential and Integral Equations
85 V. T. Fomenko et al., Twelve Papers on Functional Analysis and Geometry

(See the AMS catalog for earlier titles)